T0258888

Long-Term Stewardship and the Nuclear Weapons Complex: The Challenge Ahead

by Katherine N. Probst
and Michael H. McGovern

Center for Risk Management • Resources for the Future • June 1998

Published 1998 by Resources for the Future
2 Park Square, Milton Park, Abingdon, Oxfordshire OX14 4RN
711 Third Avenue, New York, NY 10017

First issued in hardback 2017

Library of Congress Cataloging-in-Publication Data

Probst, Katherine, N.

    Long-term stewardship and the nuclear weapons complex: The challenge ahead /
by Katherine N. Probst and Michael H. McGovern.
      p. cm.
    Includes bibliographical references.
    ISBN 0–915707–97–7 (pbk.)
    1. Nuclear weapons plants--Environmental aspects--United States. 2. Cleanup of
radioactive waste sites--United States. 3. Hazardous wastes--United States--
Management. 4. Radioactive wastes--United States--Management.
    I. McGovern, Michael H., 1965- . II. Title.

TD195.N85P77   1998
363.72'897'0973--DC21   98-23013
                                     CIP
ISBN 13: 978-1-138-16451-2 (hbk)
ISBN 13: 978-0-915707-97-3 (pbk)

Contents

Acknowledgments

The research in this report was made possible in large part by funds from the U.S. Department of Energy's (DOE) Office of Environmental Management (under agreement DE-FC01-95EW11960) to Resources for the Future's (RFF) Center for Risk Management. Without DOE's financial support, we would not have been able to undertake this work. In addition, general support from RFF's endowment and corporate contributions helped support this work.

Throughout our research, we have been fortunate to work with a great team of people at DOE. Jim Werner, Steve Livingstone, and Jenny Craig (now with the U.S. Environmental Protection Agency) have provided constructive suggestions at all stages of our work, as have many other DOE staff. We also wish to thank the numerous people—too many to name here—who provided constructive comments on a preliminary draft of this report, as well as those who attended our 1997 and 1998 workshops on long-term stewardship.

We could not have issued this report without the help of many people at RFF. Terry Davies provided useful counsel, as always. Adam Lowe and Kieran McCarthy provided excellent research assistance. Chris Kelaher and Eric Wurzbacher provided their usual high-caliber publications assistance. John Mankin typed much of the manuscript. Marilyn Voigt helped the trains run on time, as usual.

The views expressed in this report are those of the authors and should not be ascribed to the persons or organizations whose assistance is acknowledged above or to the trustees, officers, or other staff members of Resources for the Future.

Executive Summary

No matter how much money is spent to address radioactive and hazardous contamination at the sites owned and operated by the U.S. Department of Energy (DOE), some hazards will remain at most of the nation's former nuclear weapons production sites. To protect human health and the environment for current and future generations will require long-term stewardship—a program to assure protection of public health and the environment in the decades to come.

What Is the Problem?

For nearly five decades, DOE and its predecessors engaged in a highly secretive, complex, and massive endeavor to fabricate nuclear weapons. This effort required enormous facilities, material and energy inputs, and human labor. The "weapons complex" consisted of nuclear defense, nuclear energy, and research installations. These facilities were scattered across the country at large federal reservations and at smaller commercial sites. Some of these facilities housed nuclear weapons research, production, and testing activities. Others focused on civilian nuclear energy research and development activities. Huge laboratories were dedicated to nuclear research.

In the rush to produce the materials, components, and devices necessary to manufacture thousands of nuclear weapons, DOE paid scant attention to the environmental consequences of its actions. Waste materials from research and production activities were often buried on-site in shallow earth trenches or placed in settling ponds. At many sites, tremendous volumes of soil and groundwater were contaminated with hazardous and radioactive substances. Large volumes of poorly managed wastes leaked from damaged containment structures, and many aging facilities harboring highly radioactive materials deteriorated. For years, there was little information publicly available about these problems and little external regulation of DOE's environmental management activities.

With the winding down of the Cold War in the late 1980s, weapons production operations ceased. DOE turned its attention to the growing health, safety, and environmental concerns linked to past nuclear weapons production activities. This was, in large part, the result of successful litigation directed at ending DOE's immunity from federal environmental enforcement and of increased media attention. Now, almost fifteen years later, one-third of DOE's budget goes to its Office of Environmental Management (EM). At approximately $6 billion, the annual EM budget is twice as large as total estimated public and private expenditures on nonfederal Superfund sites, and just $1 billion less than the budget for the entire U.S. Environmental Protection Agency (EPA).

DOE's Office of Environmental Management is faced with the Herculean challenge of cleaning up contamination, wastes, nuclear materials, and contaminated structures at over one hundred sites in thirty states around the country. It will take decades before the department completes "cleanup" activities at all the sites in the weapons complex. The total price tag has been estimated to be somewhere between $150 and $200 billion, with most of this money going to five sites: Hanford, Savannah River, Rocky Flats, Oak Ridge, and the Idaho National Engineering and Environmental Laboratory.

Describing the EM program as a "cleanup" program is, however, something of a misnomer. No matter how much money is spent, some hazards will remain at over two-thirds of the sites. The lack of proven technologies to address radioactive contamination, and contaminated soil and groundwater, as well as the fact that many DOE sites will be home to waste storage and disposal facilities, ensures that hazards will remain at these sites for hundreds, if not thousands, of years. DOE will not be able to walk away from these sites, nor from its past contamination problems. A program of *long-term stewardship* will be needed at the majority of the sites in the weapons complex.

What Is "Long-Term Stewardship"?

Broadly speaking, stewardship refers to physical controls, institutions, information, and other mechanisms needed to ensure protection of people and the environment, both in the short and the long term, after the cleanup of the weapons complex is considered "complete." The likely elements of a stewardship program are

• Site monitoring and maintenance;
• Application and enforcement of institutional controls;
• Information management; and
• Environmental monitoring.

The notion of stewardship carries with it something more, however, than simply a list of tasks or functions to be implemented. It connotes a sacred responsibility to protect human health and the environment for future generations.

While defining the mission of a long-term stewardship program is simple, actually *creating* an enduring stewardship program is a much more difficult task. Most daunting will be to ensure that the institution (or institutions) charged with stewardship responsibilities has the bureaucratic, political, and financial wherewithal to successfully implement them.

Who Is Responsible?

It is critical that DOE begin now to consider how today's policy decisions will affect tomorrow's stewardship needs. Although DOE is already implementing stewardship activities at some of the smaller sites in the weapons complex, the department needs to take action now to show leadership on this issue and to start to lay the groundwork for developing a comprehensive stewardship program for all DOE sites.

One key issue, however, is whether DOE should continue to have a major role in stewardship of its sites, or whether responsibility for long-term stewardship should be transferred to another federal agency, or to state agencies, for certain sites. If DOE continues to have responsibility for stewardship activities at its sites, increased external oversight—by EPA, states, or some other agency—will be needed to hold DOE accountable and increase public confidence that important post-closure activities are, in fact, being implemented.

A second key issue is the scope of a stewardship program. Many contaminated sites across the United States will require post-closure care. A central question is whether a stewardship program should address all contaminated sites—whether public or private—all federal facilities subject to Superfund, or only DOE sites. The decision about the scope of a stewardship program has important implications for what organization (or organizations) should implement stewardship functions, what institution should be responsible for stewardship oversight, and, finally, for how a stewardship program should be created and funded.

EPA, too, bears an important responsibility for addressing these issues. Superfund is one of the primary statutes driving cleanup activities at DOE and many other contaminated sites. The increasingly frequent use of institutional controls as an integral component of site remedies and their potential application at DOE sites demands that the issue of assuring the long-term integrity of institutional controls be addressed. This could be accomplished by amending the Superfund law, or by revising the major Superfund regulation, the National Contingency Plan. In fact, the Superfund reauthorization debate may well provide the best opportunity for creating a legislative stewardship mandate. After all, stewardship activities will be required not only at DOE sites, but at many Superfund sites, including those contaminated sites under the purview of other federal agencies, such as the Departments of Defense, Agriculture, and Interior. Moreover, because some of the largest, most complicated, and most expensive DOE sites are on EPA's National Priorities List, creating a stewardship program under Superfund has the advantage of addressing both the major DOE sites and other contaminated sites.

We have concluded that the primary locus for stewardship should be the federal government. It is one of the most enduring of today's institutions, other than religious entities. Federal agencies should (at least initially) have primary responsibility for stewardship at DOE sites, with the stewardship mission, goals, and objectives set out by Congress in federal authorizing legislation. Equally important, federal appropriations specifically earmarked for stewardship activities will be needed, both to fund the program and to confirm the federal government's commitment to long-term stewardship.

While some federal role in stewardship will be necessary, states, localities, tribal nations, and the general public must be meaningfully involved in the development, implementation, and oversight of stewardship activities. The involvement of stakeholders in stewardship will help increase public trust in a stewardship program and ensure much-needed external accountability. History suggests that the involvement of these other entities is critical to keeping the federal system "honest."

What Do We Do Next?

The most important "next step" is to stimulate a public dialogue about the need for long-term stewardship at the nuclear weapons complex sites, and about the appropriate scope of a stewardship program. Addressing these issues is the responsibility of all stakeholders—DOE, EPA, state agencies, local governments, tribal nations, citizens' groups, and private industry. Absent their involvement and support, it will be difficult to take the steps needed to create an effective and credible stewardship program.

We recommend two kinds of actions that should be taken to move this important issue forward:

- Steps that can be taken now to create a stewardship program, and
- Research and analysis that should be conducted to help support the development of a stewardship program.

Both are outlined below. Needless to say, there should be early, meaningful, and continual involvement of all stakeholders in these activities.

Creating a Stewardship Program

In order to assure the creation of a successful stewardship program, a legislative or regulatory stewardship mandate is needed. Barring that, some sort of administrative action should be taken. The recommendations below are listed in priority order.

1. Congress should enact stand-alone stewardship legislation or amend Superfund and the Resource Conservation and Recovery Act (RCRA) to include specific language requiring the creation of a stewardship program for all contaminated sites requiring post-closure care that are regulated under the nation's environmental laws.

2. EPA should amend the National Contingency Plan to clearly define post-closure responsibilities at Superfund sites on the part of federal, state, and local governments, and regulated entities—both public and private.

3. The President's Council of Environmental Quality, jointly with EPA, should convene an interagency task force that also includes independent experts and representatives of major stakeholders to develop a government-wide policy on long-term stewardship at both federal and private sites regulated under Superfund and RCRA.

4. The secretary of DOE should create a high-level task force that includes independent experts from the full panoply of stakeholders—federal, state, and local governments; tribal nations; environmental groups; private industry; and academia—to develop a stewardship mission for DOE, and to make specific recommendations for integrating the costs and the challenges of long-term stewardship into the major DOE internal decision making and budgeting processes.

Stewardship Research and Analysis

Whether a mandate for stewardship is created now, or years from now, better information on the scope, nature, and cost of stewardship is needed.

1. EPA should commission two studies regarding stewardship at both federal and private sites: (a) an examination of the role of states and local governments as stewardship implementors and/or overseers, and (b) an evaluation of a range of institutional alternatives for assuring long-term compliance with institutional controls.

2. EPA should assess the need for long-term stewardship at sites addressed under Superfund and Subtitle C of RCRA, and estimate the full cost of stewardship to both the public and private sector at these sites.

3. DOE should (a) estimate the cost, time frames, and types of activities that will be needed for long-term stewardship at the sites in the weapons complex; and (b) commission an independent report regarding the appropriate role of tribal nations and local governments in long-term stewardship at DOE sites.

4. The Congressional Budget Office should conduct a study of alternative funding schemes, for both federal and private sites, for paying for long-term stewardship.

5. The Congressional Research Service or National Academy of Sciences should conduct a study of how other countries are meeting their stewardship responsibilities for post-closure care at contaminated sites to help inform the development of a U.S. program.

I. Introduction

The U.S. Department of Energy (DOE) is faced with the Herculean challenge of cleaning up the contamination, wastes, nuclear materials, and contaminated structures resulting from decades of nuclear weapons production and nuclear energy research and development at over one hundred sites in thirty states around the country.[1] Most experts believe that it will take decades before the department completes "cleanup" activities at all the sites in the weapons complex. The total price tag has been estimated to be somewhere between $150 and $200 billion.[2] Most of this money will be spent at five sites: Hanford, Savannah River, Rocky Flats, Oak Ridge, and the Idaho National Engineering and Environmental Laboratory.

Even though DOE's official goal is to "clean up" the sites in the nuclear weapons complex, no matter how much money DOE spends, some hazards will remain at

---

[1]Until recently the official number of sites was over 130. With the recent transfer of authority for the Formerly Used Site Remedial Action Program from DOE to the Army Corps of Engineers, that number has decreased to 113 sites.

[2]See Office of Environmental Management, *Accelerating Cleanup: Paths to Closure,* Draft, DOE/EM-0342 (Washington, D.C.: U.S. Department of Energy, February 1998); and Office of Environmental Management, *The 1996 Baseline Environmental Management Report,* DOE/EM-0290 (Washington, D.C.: U.S. Department of Energy, June 1996).

over two-thirds of these sites.[3] This is due, in large part, to the nature of the con-tamination, and the lack of proven cleanup and treatment technologies. Thus, referring to DOE's efforts as a "cleanup" program is somewhat of a misnomer. In fact, hazards remaining at some DOE sites will require attention for many cen-turies to come. Land use will need to be restricted for portions of some sites, as will the future use of contaminated groundwater.

The decisions made today about what problems to address, how much to clean up, and how and where to dispose of contaminated materials and haz-ardous and radioactive wastes will affect the distribution and degree of risk that remain at DOE sites. Decisions made now about the future use of nuclear research and weapons production sites will also affect the nature of the legacy DOE leaves for future generations. These choices will determine the kinds of long-term government responsibilities that will be necessary to ensure protec-tion of human health and the environment in the future—that is, the need for long-term stewardship, the subject of this report. Broadly speaking, stewardship refers to physical controls, institutions, information, and other mechanisms needed to ensure protection of people and the environment, in both the short and the long term, after the cleanup of the weapons complex is considered "complete." The notion of stewardship carries with it something more, however, than a list of tasks or functions to be implemented. It connotes a sacred respon-sibility to protect human health and the environment for future generations.

There are two major types of challenges to developing and implementing a successful stewardship program: technical and institutional. The technical challenges—primarily the fact that there is currently no proven technology that can render radioactive materials harmless—are not really subject to quick gov-ernment fixes. This technology gap is further complicated by the fact that it will take hundreds, and even thousands, of years for long-lived radioactive contam-inants at DOE sites to fully decay to background levels. It is largely because of this technical challenge that a stewardship program is needed.

The institutional challenges are equally daunting. What organization or organizations should be charged with ensuring protection at these sites? Under what legislative authority should a stewardship organization operate? What is the appropriate role for state and local governments, tribal nations, and other stakeholders? And, perhaps most difficult, how can the long-term financial secu-

---

[3]Office of Environmental Management, *Moving from Cleanup to Stewardship*, Working Draft (Washington, D.C.: U.S. Department of Energy, September 17, 1997): 15–17.

rity of the organization charged with stewardship be ensured? These are difficult questions to answer, given the political context surrounding the weapons complex and the long-lived nature of the hazards involved.

Not surprisingly, the need to think about how to ensure long-term protection of human health and the environment from long-lasting hazards is not unique to DOE sites and facilities. In fact, a number of existing federal regulatory programs already contain elements of a stewardship program. Some of these programs apply to DOE sites, others do not. In general, these programs concern either contaminated sites and facilities or waste disposal sites and facilities. Typically, the stewardship functions (although they are not referred to as such) relate to what is called "post-closure" care, and consist of steps to ensure protection after a facility has ceased operation or after remediation at a site has been completed.

A list of some, but not all, of the programs that include aspects of what we are calling long-term stewardship follows:[4]

- *Hazardous waste land disposal* under the Resource Conservation and Recovery Act
- *Cleanup of sites contaminated with hazardous substances*, under Superfund (the Comprehensive Environmental Response, Compensation, and Liability Act)
- *Disposal of low-level radioactive waste generated by nuclear power facilities and other sources*, under the Nuclear Waste Policy Act and Atomic Energy Act
- *Decommissioned licensed nuclear power facilities*, under the Atomic Energy Act
- *Cleanup and disposal of uranium mill tailings*, under the Uranium Mill Tailings Remedial Control Act

These programs are in varying stages of implementation. Most of these stewardship activities are still in their infancy, and thus have not yet been evaluated. Still, they show that the need for long-term monitoring, maintenance, and institutional controls is not unique to DOE sites, and that precedents exist for regulatory programs requiring long-term controls to ensure protection of human health and the environment. Indeed, the stewardship functions that we propose in Section III stem in part from our review of the previously listed programs.

---

[4]See Appendix A for more information on selected federal programs with stewardship elements.

It is critical that DOE begin now to consider how today's policy decisions will affect tomorrow's stewardship needs. Even more importantly, DOE and the full panoply of stakeholders involved at the weapons sites—the U.S. Environmental Protection Agency (EPA), state agencies, local citizens, tribal nations, local governments, environmental groups, DOE contractors—must begin now to develop a framework for ensuring a successful program of long-term stewardship. Congress and the agencies that regulate DOE (EPA, other federal agencies, and states) must also begin to wrestle with these difficult issues. Equally important will be designating an institution—or institutions—to carry out and oversee a program of long-term stewardship at many of the sites in the weapons complex.

The purpose of this paper is to stimulate discussion about the need for long-term stewardship at the sites in the nuclear weapons complex. The report is structured to take the reader through the major components of an argument for why such a program is necessary. First, we describe the environmental and institutional legacy of decades of weapons production. Second, we lay out key functions of a long-term stewardship program. Third, we identify important institutional issues that must be addressed to develop a successful stewardship program, and discuss the pros and cons of several institutional alternatives for carrying out stewardship activities. Finally, we present our recommendations for moving forward to address the challenge of long-term stewardship at the sites in the nuclear weapons complex.

II. The Legacy of the Cold War

For nearly five decades, the U.S. Department of Energy (DOE) and its predecessors engaged in a highly secretive, complex, and massive endeavor to fabricate nuclear weapons for national security purposes.[5] The production of nuclear weapons required enormous facilities, material and energy inputs, and human labor. The "weapons complex" consisted of nuclear defense, nuclear energy, and research installations. These facilities were scattered across the country at large federal reservations and at smaller commercial sites. Some of the facilities housed nuclear weapons research, production, and testing activities. Others focused on civilian nuclear energy research and development activities. Huge laboratories were dedicated to nuclear research. These activities involved a broad range of uses of nuclear materials, including uranium mining and milling, uranium enrichment, reactor operations, and other activities. Many of the materials and wastes that were used or generated as part of the weapons production process remain highly dangerous to human health and the environment.

In the rush to produce the materials, components, and devices necessary to manufacture thousands of nuclear weapons, DOE paid scant attention to the environmental consequences of its actions. Waste materials from research and production activities were often buried

---

[5]For ease of reference, throughout this report we refer to DOE and its predecessor agencies as "DOE."

5

on-site in shallow earth trenches or placed in settling ponds. At many sites, tremendous volumes of soil and groundwater were contaminated with hazardous and radioactive substances. Large volumes of poorly managed wastes leaked from damaged containment structures, and many aging facilities harboring highly radioactive materials deteriorated.[6] For years, there was little information publicly available about these problems and little external regulation of DOE's environmental management activities.[7]

With the winding down of the Cold War in the late 1980s, weapons production operations ceased. DOE turned its attention to the growing health, safety, and environmental concerns linked to past nuclear weapons production activities. This was, in large part, the result of successful litigation directed at ending DOE's immunity from federal environmental regulation[8] and of increased media attention.[9] The appointment of Hazel O'Leary as DOE secretary in 1993 ushered in a new, more open, era. The veil of secrecy that for so long shielded the nuclear weapons production complex from outside scrutiny and external regulation began to lift.

In 1989, then Secretary of Energy James Watkins established the Office of Environmental Restoration and Waste Management, subsequently renamed Environmental Management or "EM," to clean up the sites of the nuclear weapons complex. This marked a major change in the structure and mission of

---

[6]For an excellent overview of the environmental problems stemming from nuclear weapons production, see Office of Environmental Management, *Closing the Circle on the Splitting of the Atom* (Washington, D.C.: U.S. Department of Energy, January 1996). For a more detailed analysis, see Office of Environmental Management, *Linking Legacies: Connecting the Cold War Nuclear Weapons Production Processes to Their Environmental Consequences* (Washington, D.C.: U.S. Department of Energy, January 1997).

[7]For a discussion of DOE's regulatory framework, see Advisory Committee on External Regulation of Department of Energy Nuclear Safety, *Improving Regulation of Safety at DOE Nuclear Facilities* (Washington, D.C.: U.S. Department of Energy, December 1995); and U.S. Department of Energy, *Report of Department of Energy Working Group on External Regulation* (Washington, D.C., December 1996).

[8]*Legal Environmental Assistance Foundation v Hodel*, 586 F Supp 1163 (ED Tenn D Ct 1984); *Natural Resources Defense Council (NRDC) v EPA*, 824 F2nd 1258 (1st Cir 1987); and *NRDC v DOE*, Cir Action No. 89-1835 (D DC 1989).

[9]William Lanouette, *Tritium and the Times: How the Nuclear Weapons-Production Scandal Became a National Story*, Research Paper R-1 (Cambridge, Mass.: Joan Shorenstein Barone Center on the Press, Politics, and Public Policy, Harvard University, May 1990).

the Department of Energy. Today, EM's budget is the largest of the offices within DOE, comprising one-third of DOE's total annual appropriations.[10] At $6 billion a year, EM has one of the largest environmental management budgets of any federal agency in the United States, and perhaps, in the world.[11]

Much of EM's budget goes to activities few would define as "environmental management." In fact, EM estimates that in recent years as much as 50% of its budget has gone to what are referred to as the "mortgage" costs of ensuring security and maintaining local infrastructure at the sites in the weapons complex.[12] Some funds also go to managing excess nuclear materials stored at DOE sites. Since 1989, DOE has spent over $40 billion on this diverse set of tasks.

DOE is subject to an array of environmental regulations, primarily under the Resource Conservation and Recovery Act (RCRA) and the Comprehensive Environmental Response, Compensation, and Liability Act (CERCLA, better known as Superfund).[13] RCRA regulations govern the management of hazardous wastes, including the portion of mixed wastes (wastes that include both radioactive and hazardous wastes) that are hazardous. However, the U.S. Environmental Protection Agency (EPA) does not have authority to regulate management and disposal of radioactive wastes under RCRA. Most DOE sites are subject to state regulation under RCRA as well.[14] Nuclear wastes and materials generated by DOE are regulated by DOE, under the Atomic Energy Act (AEA).[15] This regulation by DOE of its own activities is referred to as "self-regulation." DOE must comply with a host of internal directives and regulations

---

[10]U.S. Department of Energy, "FY 1999 Control Table by Organization," in *FY 1999 Department of Energy Budget Request* (Washington, D.C., February 1998).

[11]The total budget of the U.S. EPA is larger, at $7 billion annually, but DOE has the largest budget of any federal agency for *complying* with the nation's environmental laws.

[12]Testimony of Alvin L. Alm, Assistant Secretary for Environmental Management, as recorded in U.S. House Committee on Appropriations, *Energy and Water Development Appropriations for 1998: Hearing before the Subcommittee on Energy and Water Development,* 105th Cong., 1st Sess. (March 12, 1997).

[13]*Resource Conservation and Recovery Act of 1976,* Public Law 580, 94th Cong., 2nd Sess. (October 21, 1976); and *Comprehensive Environmental Response, Compensation, and Liability Act of 1980,* Public Law 510, 96th Cong., 2nd Sess. (December 11, 1980).

[14]Facilities at DOE sites must also comply with appropriate provisions of the other major environmental statutes, such as the Clean Air Act, the Clean Water Act, and the National Environmental Policy Act.

[15]*Atomic Energy Act of 1954,* Public Law 703, 83rd Cong., 2nd Sess. (August 30, 1954).

that implement the provisions of the AEA and govern the management and storage of nuclear materials.

In 1986, Congress amended CERCLA to make clear that federal facilities are subject to Superfund.[16] Unlike RCRA, Superfund gives EPA authority to address radioactive substances. Fifteen DOE sites are on the EPA's Superfund National Priorities List (NPL) and thus must be cleaned up in compliance with Superfund regulations.[17] Typically, DOE cleanup activities are governed by legally binding agreements signed by DOE, EPA, and the relevant state environmental agency. At some sites, DOE also enters into agreements with tribal governments. In addition, cleanups of many other contaminated DOE sites (that is, those that are not on the NPL) are conducted in accordance with Superfund policies and regulations.

Another key piece of legislation for DOE sites is the Federal Facility Compliance Act (FFCA), which was enacted in 1992.[18] By expressly waiving federal sovereign immunity for violations of RCRA, the FFCA made it clear that federal facilities are subject to penalties. Thus, the FFCA gave EPA explicit authority to enforce environmental regulations at DOE sites and at other federal facilities. In addition, the FFCA required DOE to submit to EPA plans by a certain date for how it intended to manage and dispose of its mixed wastes.

The Environmental Legacy: Intractable Problems

The environmental legacy of nuclear weapons production includes diverse kinds of hazards. Some of these hazards are typical of pollution problems at many industrial facilities: hazardous materials and wastes that require proper storage and disposal, and soil and groundwater contaminated with solvents, oils, and other chemicals. Other hazards are unique to the weapons complex: nuclear materials that require special handling, large volumes of radioactive waste, and deteriorating facilities, many of which are themselves contaminated with radioactive and hazardous substances.

---

[16]*Comprehensive Environmental Response, Compensation, and Liability Act,* § 120.

[17]U.S. Department of Energy, "Table ES-1. U.S. Department of Energy Facilities on the National Priorities List," in *FY 1996 Progress in Implementing Section 120 of CERCLA: Tenth Annual Report to Congress,* DOE-EM-0330 (Washington, D.C., December 1997).

[18]*Federal Facility Compliance Act of 1992,* Public Law 386, 102nd Cong., 2nd Sess. (October 6, 1992).

The Office of Environmental Management (EM) is responsible for waste management, environmental restoration, facility decontamination and decommissioning, and nuclear materials management. EM activities are currently underway at fifty-three sites. These sites include[19]

- Fourteen nuclear weapons production sites—the five "major" sites previously noted (Hanford, Savannah River, Rocky Flats, Oak Ridge, and the Idaho National Engineering and Environmental Laboratory [INEEL]) and nine other sites that were involved in the fabrication of nuclear weapons components or production of weapons-grade uranium;
- Ten sites that were used for testing of nuclear bombs and other explosives;
- Nineteen DOE research laboratories—fourteen of which continue to have active DOE research missions;
- Five uranium mill processing sites; and
- Five other sites in the EM program—the future transuranic waste disposal facility, two private facilities that engaged in nuclear energy research, a closed commercial low-level waste disposal site, and a former commercial nuclear fuel reprocessing plant.

EM is also responsible for another 60 sites, that have been deemed "complete" by EM, although some long-term stewardship activities will be required at many of these sites.[20]

Some EM facilities are vast (INEEL is 890 square miles; the Nevada Test Site is 1,350 square miles); others are quite small (the Laboratory for Energy-Related Health Research is fifteen acres). A number of these sites are of tremendous importance to tribal nations that have lived near these sites for centuries. In addition, parts of some sites have been declared national environmental research parks because they house unspoiled ecosystems and some endangered species.[21]

The EM program is responsible for managing thirty-six million cubic meters of "legacy" wastes. That is enough waste to bury the entire island of Manhattan in two-and-a-half feet of waste. These are wastes resulting from Cold War activities, two-thirds of which are the result of nuclear weapons production activities. Legacy wastes include high-level, transuranic, low-level, mixed (radio-

---

[19]See Appendix B for a list of these fifty-three sites and their associated EM budget.

[20]Office of Environmental Management, *Accelerating Cleanup: Paths to Closure*, p. 8.

[21]See Virginia H. Dale and Patricia D. Parr, "Preserving DOE's Research Parks," *Issues in Science and Technology* (Winter 1997–98): 73–77.

active and hazardous), and hazardous wastes; as well as uranium and thorium mill tailings and wastes. Mill tailings represent the majority (thirty-two million cubic meters) of the waste volume, though they contain comparatively low levels of radioactivity.[22]

ENVIRONMENTAL LEGACY

Total wastes	**36 million cubic meters**
Uranium mill tailings	32 million cubic meters
Low-level wastes	3.3 million cubic meters
High-level wastes	380,000 cubic meters
Transuranic wastes	220,000 cubic meters
Mixed low-level and other wastes	215,000 cubic meters
Contaminated soils	**75 million cubic meters**
Contaminated groundwater	**1.8 billion cubic meters**
Excess facilities (already in EM program)	**10,000 buildings and structures**

Sources: Office of Environmental Management, *Linking Legacies,* pp. 58, 80; and Office of Environmental Management, *Accelerating Cleanup: Focus on 2006,* Discussion Draft, DOE/EM-0342 (Washington, D.C., June 1997): E-1.

There is currently no operating permanent disposal facility for high-level radioactive or transuranic wastes, although the Waste Isolation Pilot Plant is scheduled to open soon. In addition to legacy wastes, DOE facilities continue to generate new wastes as part of ongoing missions. These wastes, too, must be properly managed. Finding adequate and politically acceptable waste disposal capacity is one of the major challenges facing DOE.

In addition, enormous quantities of soils, sediments, groundwater, and surface water are contaminated with hazardous and radioactive substances as a result of nuclear weapons production, research, and testing. DOE estimates that it is responsible for over 75 million cubic meters of contaminated soil and approximately 1.8 billion cubic meters of contaminated groundwater, or enough to flood Manhattan in 135 feet of water.[23] The dearth of effective treatment technologies for contaminated soil and groundwater means that many of these problems will endure for many decades to come.

---

[22]Office of Environmental Management, *Linking Legacies,* p. 58.

[23]Ibid., p. 80.

DOE's environmental management program is also responsible for decontaminating and decommissioning 10,000 "excess" facilities. Many of these buildings and structures are contaminated with hazardous and radioactive substances. Some of these facilities are huge, as long as skyscrapers lying flat; others are small shacks. In the near future, an additional 800–1,500 facilities could be declared "excess," one-third of which are estimated to be contaminated with hazardous and/or radioactive substances.[24]

Finally, DOE sites house large surplus inventories of materials used to manufacture nuclear warheads, notably, plutonium and highly enriched uranium. Current DOE inventories include large volumes of spent nuclear fuel, natural uranium, depleted uranium, lithium, sodium, lead, various other chemicals, nonnuclear weapons components, and scrap metals. Current plans call for most of these excess materials to be consolidated, sold, reused, or disposed of. Materials stored or disposed of on-site are potential hazards that have stewardship implications. Addressing these hazards is complicated by the need to maintain these materials in secure locations. Plutonium disposition is also complicated by debates about whether surplus plutonium should be immobilized and then disposed of as waste, or whether it should be converted into reactor fuel, burned, and disposed of as spent fuel, or both.

All of these legacies of the Cold War present some risks to human health and the environment—either now or in the future. The most obvious are those resulting from the presence of radioactive and hazardous substances. However, at many sites, the largest current risks arise from the deteriorating physical state of buildings, facilities, and equipment. These deteriorating structures can result in high risks to on-site workers. Failure to properly maintain buildings has had tragic consequences: in one incident, a worker was killed when he fell through a roof in poor repair at Hanford.

The EM program is responsible for getting legacy wastes, contaminated media, surplus buildings, and nuclear materials cleaned up to their "end state." According to DOE, the end state is achieved (that is, EM activities are deemed complete) when

- "Legacy" waste (that is, the waste produced by past nuclear weapons production activities) has been disposed of in an approved manner;

---

[24]Office of Environmental Management, *Accelerating Cleanup: Focus on 2006*, Discussion Draft, DOE/EM-0342 (Washington, D.C., June 1997) E-1, E-2.

- Groundwater contamination has been contained, and long-term treatment or monitoring is in place;
- All releases to the environment have been cleaned up in accordance with the agreed-upon standards;
- Deactivation or decommissioning of all facilities currently in the EM program has been completed, excluding any long-term surveillance and monitoring; and
- Nuclear material and spent fuel have been stabilized and/or placed in safe long-term storage.[25]

Much of the contamination, wastes, facilities, and materials in inventory at DOE sites will be treated, managed, or disposed of on-site.[26] Thus, hazards will remain—albeit in a more stable form—at many of the sites in the weapons complex. The problems likely to remain at DOE sites include

- Extensive on-site groundwater and soil contamination, as well as some off-site contamination;
- Buried trenches and ponds containing nuclear and hazardous materials or wastes;
- Underground waste disposal facilities in which hazardous waste, low-level waste, mixed radioactive waste, and uranium mill tailings will be buried in large volumes;
- Large contaminated facilities, either entombed in concrete and buried in earthen mounds, or collapsed and buried in place; and
- Long-term storage facilities that will contain dangerous nuclear materials and/or radioactive wastes that cannot be quickly disposed of.

What happens next—that is, who is responsible for ongoing monitoring and maintenance of facilities and institutional controls, or what we call "stewardship"—is not clear. It is unlikely that planned disposal and containment practices will be capable of containing long-lasting radioactive materials, wastes, and contamination for the time period over which some of these hazards will remain harmful—hundreds if not thousands of years. (See the inset for information about radioactive decay.) Even in the short term (decades), it is less than certain

---

[25]Office of Environmental Management, *Accelerating Cleanup: Paths to Closure*, p. 19.

[26]The big exceptions are spent nuclear fuel and high-level waste, which are scheduled to go to a geological repository (probably Yucca Mountain) at some time in the future; transuranic waste, most of which is supposed to go the Waste Isolation Pilot Plant when it opens; highly enriched uranium; and plutonium.

that all of the containment mechanisms employed at sites will be effective. This is in large part because so-called engineered barriers are likely to deteriorate over time. Natural phenomena—such as hurricanes, tornadoes, earthquakes, burrowing animals, floods, and groundwater movement—can hamper their effectiveness, leading to the dispersal of materials contained within. In addition, the activity of human beings at and below the surface could penetrate engineered barriers and compromise their integrity. These activities include exploring for minerals or oil, building foundations for buildings, drilling drinking-water wells, and installing pipelines and cables for electricity or communications.

RADIOACTIVE DECAY

Radioactive substances decay at a fixed rate, largely unaffected by temperature, solvents, or seasons of the year. The rate of decay is measured by the "half life"—the amount of time required for one-half of a given amount of a radionuclide to decay.

Radioactive contaminants have the potential to present some hazard for about ten times the half-life of a given isotope. Tritium, for example, with a rel-

Half-lives of Typical Constituents in DOE Waste and Contamination	
Tritium	12 years
Plutonium 241	14 years
Plutonium 238	89 years
Radium 226	1,600 years
Plutonium 240	6,570 years
Plutonium 239	24,100 years
Thorium 230	80,000 years
Uranium 235	703,800,000 years
Uranium 238	4,468,000,000 years

Source: Moving from Cleanup, p. 5.

atively short half-life of 12 years, remains a potential hazard for approximately 120 years. Plutonium 239, with a half-life of 24,100 years, will present a hazard for nearly a quarter million years.

Given the scope of potential risks that will remain at many sites, it is clear that a follow-up program of some kind—a program of long-term stewardship—will be necessary to ensure that today's cleanup measures remain effective in both the short and long term. The required scope and associated costs of a stewardship program will depend substantially on what specific remedial strategies are implemented at each of the DOE sites. At least in theory, the more thorough today's cleanup solutions—that is, the more permanent the cleanup solutions implemented and the lower the dependence on containment strategies and institutional controls—the fewer demands placed on a future program of long-term stewardship. Of course, there are limits in terms of how much "permanence" can be achieved, because of both technological and financial constraints. Less stringent cleanup solutions will result in a greater need for long-term stewardship.

The Institutional Legacy: Multiple Missions and a Culture of Secrecy

Two things stand out about the way DOE and its predecessor agencies operated during the Cold War: their focus on secrecy and their single-minded determination to build bombs. For decades, these organizational mores guided the plans and actions of agencies charged with weapons production. To ensure that national security was maintained while weapons production was underway, these agencies purposefully created compartmentalized organizational structures, kept secret from the public certain production facilities, and minimized communication among facilities. While such measures were advantageous from a security standpoint, they ultimately created an unwieldy and expensive bureaucracy that has appeared at times to be accountable to no one.

Over the past decade, efforts to change the organization and culture of the department have met with mixed results. Under the Bush administration, Secretary Watkins took several positive steps toward addressing the environmental problems of the weapons complex. For example, in addition to establishing the Office of Environmental Management, he put in place investigative "tiger teams" to perform comprehensive environment, safety, and health assessments at major DOE facilities. Under the Clinton administration, Hazel O'Leary also set in motion major changes in the DOE institutional culture. These included the Openness Initiative, a process that has led to the declassification of an enormous number of secret documents, and a transition from self-regulation to external regulation.[27] Both Secretary Watkins and Secretary O'Leary exerted considerable pressure on the DOE bureaucracy to try to achieve "cultural change" within the department.

---

[27]In response to the recommendations of an independent panel convened by former DOE Secretary O'Leary (the "External Regulation" or "Ahearne Commission"), DOE recommended that regulation of nuclear safety at DOE facilities be given to the U.S. Nuclear Regulatory Commission.

Despite this considerable progress, institutional problems within the department persist. Every few months, a new story appears in the national papers about environmental problems not disclosed, or an environmental management project gone awry. One recent example is the admission by DOE in March of 1998 that it made a mistake when it assured the public that radioactive wastes leaking into the ground at Hanford would not contaminate groundwater. This was based on the faulty assumption (as it turns out) that liquid waste would not move very far in the region of the soil called the "vadose zone." To compound DOE's embarrassment, it appears that DOE did not put a lot of effort into trying to understand how plutonium moves through soil, even though this is a major concern at Hanford.[28]

Also in March of 1998, DOE fined one of its Hanford consultants, Fluor Daniel Hanford, Inc., $140,625 for nuclear safety violations that included a release of small amounts of radioactive plutonium into the environment.[29] Meanwhile, the state of Washington threatened to sue DOE Secretary Federico Peña because of missed cleanup deadlines at Hanford, yet again.[30]

While these and other examples can be construed as simply anecdotes, these stories leave some wondering whether DOE has really embraced its new mission of environmental management, and whether it is exerting adequate management leadership of its own operations and its contractors, who do most of the work at DOE field offices. In many ways, DOE continues to operate as an autonomous institution, without adequate external oversight by EPA or other federal agencies.[31]

The Cold War left a complicated institutional legacy. DOE is in many ways a conglomeration of independent program offices—environmental man-

---

[28]See Matthew L. Wald, "Admitting Error at a Weapons Plant; Belatedly, Energy Department Deals with Leaks of Nuclear Waste," *The New York Times* (March 23, 1998): A10; and "GAO: DOE Does Not Understand Hanford Vadose Zone," *Weapons Complex Monitor* (March 30, 1998): 3–4.

[29]"Hanford: DOE to Issue Fine for Nuclear Safety Violations," *Greenwire* (March 31, 1998); and "DOE Proposes Fines for Fluor Daniel Related to Pu Plant Explosion," *Weapons Complex Monitor* (April 6, 1998): 6.

[30]"Wash. Governor Threatens Penā [sic] with Lawsuit Over Hanford Cleanup," *Weapons Complex Monitor* (March 2, 1998): 14.

[31]For a discussion of the need for external regulation of DOE nuclear safety activities, see Andrew Caputo, "A Failed Experiment," *The Environmental Forum* (January–February 1996): 16–21.

agement, defense programs, energy research—each having its own mission and its own army of contractors and subcontractors.[32] There is little semblance of a unified mission for the department. While this situation is not unique to DOE, it has major repercussions for EM and for the department's environmental operations. For example, most current EM sites were once defense programs (DP) facilities. In recent years, DP had little incentive to control environmental contamination, as once the facility was no longer needed, it was transferred to EM. Even now, DOE sources say that many DOE program offices are not accounting for future environmental liabilities. EM has refused to accept additional facilities for two years, yet none of the other program offices have accounted for excess facility decommissioning costs in their out-year budgets.[33] This lack of cohesion is exacerbated by the fact that there is little in the way of a department-wide management or accountability structure, and no effective means to address cross-cutting issues.[34] DOE program offices operate largely independently, even as they coexist at DOE facilities around the country. Even within program offices, which carry out projects at dozens of sites across the DOE complex, there are sometimes significant differences in management and accounting from site to site. In the box on the following pages, we briefly describe the key functions of the major DOE offices that operate at DOE sites around the country. At many sites where long-term stewardship will be required, DOE program offices will continue to carry out multiple missions side by side. The ongoing missions of these program offices will have an important impact on the amount, nature, and scope of necessary stewardship activities at DOE sites. And, ongoing missions will continue to generate new wastes and other hazards, creating new stewardship demands.

---

[32]Paul H. Richanbach, David R. Graham, James P. Bell, and James D. Silk, *The Organization and Management of the Nuclear Weapons Program* (Alexandria, Va.: Institute for Defense Analyses, February 27, 1997).

[33]"DOE Struggles with Influx of New Facilities to Cleanup Program," *Inside EPA's Superfund Report* (April 1, 1998): 10; and Office of Environmental Management, *Accelerating Cleanup: Focus on 2006*, E-1, E-2.

[34]Richanbach and others, *Organization and Management;* General Accounting Office, *Department of Energy: A Framework for Restructuring DOE and Its Missions*, GAO/RCED-95-197 (Washington, D.C., August 21, 1995); and "DNFSB [Defense Facilities Nuclear Safety Board] Chair to Secretary Peña—You Must Take Control Now!" *Weapons Complex Monitor* (December 15, 1997): 2–3.

DEPARTMENT OF ENERGY PROGRAM OFFICES

The **Office of Environmental Management (EM)** is the largest DOE office, with a fiscal year 1998 budget of slightly less than $6 billion, over one-third (36%) of DOE's total appropriations. EM is charged with protecting human health and the environment from the radioactive and other hazards generated from DOE operations (and those of DOE's predecessor agencies). To address these hazards—radioactive and hazardous wastes, contaminated environmental media, contaminated facilities, and surplus materials—EM engages in a wide range of activities at 113 sites, including waste management and disposal, environmental remediation, facility decontamination and decommissioning, and nuclear materials management. EM is also responsible for what are called "landlord" functions at certain sites in its portfolio, some of which resemble small cities with their own roads, power plants, and fire departments. Landlord functions include day-to-day security operations, utility management, emergency and communications services, and basic infrastructure maintenance.

The next largest DOE office is the **Office of Defense Programs (DP)**. In FY 1998, DP's budget was just over $4 billion, one-quarter (25%) of DOE's total appropriations. Until the late 1980s, DP's role was to manufacture nuclear weapons and manage the mammoth industrial complex that produced them. Since the end of the Cold War, however, nuclear weapons production in the United States has ceased. The Office of Defense Programs is now focused on four major missions: downsizing the nuclear weapons manufacturing complex; stewarding and reducing the existing nuclear weapons stockpile; ensuring that nuclear weapons production and testing capabilities (not capacities) are preserved, should they be needed in the future; and verifying the safety and reliability of nuclear weapons without physically testing them. DP activities occur primarily at four plants, three national laboratories, and the Nevada Test Site. The sites that form the bulk of EM's mission are former DP facilities.

The **Office of Energy Research's (ER)** mission is to undertake and sponsor research in energy-related areas, including basic energy science, magnetic fusion energy, high energy and nuclear physics, health and environmental sciences, and computational science. These programs rely on a range of large research facilities, including nuclear reactors, particle accelerators, and particle detectors. ER serves as the planning, policy, oversight, and support office for the energy and multipurpose research laboratories. ER is the third largest DOE office, with a fiscal year 1998 budget of $2.5 billion, almost 15% of DOE's total appropriations. ER-supported energy research takes place at about twenty-five DOE laboratories and facilities.

The mission of the **Office of Nuclear Energy, Science, and Technology (NE)** is to provide technical leadership for domestic and international nuclear security and safety issues and to maintain nuclear energy as a viable source to meet future energy requirements and environmental objectives in the United States and abroad. Examples of NE programs include the Advanced Radioisotope Power Systems program, which provides advanced nuclear power sources to the National Aeronautics and Space Administration and national security customers; and the Naval Reactors program, which provides the Navy with specialized nuclear propulsion plants to power naval vessels. NE is the fourth largest DOE office, with a fiscal year 1998 budget of $1 billion, almost 6% of DOE's total appropriations. NE-supported projects are under way at around twenty-five DOE sites.

The **Office of Energy Efficiency and Renewable Energy (EE)** develops and promotes energy efficiency and renewable energy technologies. Research, development, and market promotion programs focus on four classes of energy users: utilities, industry, transportation, and buildings. EE also administers the Federal Energy Management Program, a program to reduce the energy consumption of federal facilities 30% by 2005. EE's fiscal year 1998 budget is $863 million, over 5% of DOE's total appropriations. Research activities occur primarily at the National Renewable Energy Lab in Colorado, although EE funding supports numerous technology demonstration projects at non-DOE facilities around the country. Six regional offices administer EE programs and services.

The **Office of Nonproliferation and National Security (NN)** is responsible for all of the Department of Energy's activities relating to nonproliferation, energy intelligence, nuclear safeguards and security, document classification and declassification policy, and emergency management. Its nonproliferation efforts include programs to prevent the spread of nuclear weapons technology worldwide, monitor compliance with arms control treaties, and secure nuclear materials in the former Soviet Union. Domestically, it safeguards and oversees the protection of nuclear materials and facilities. NN's fiscal year 1998 budget is $657 million, almost 4% of DOE's total appropriations.

The **Office of Fossil Energy (FE)** conducts and sponsors fossil fuel research and manages petroleum reserves owned by the federal government. FE's research programs include technologies for advanced power systems with reduced emissions and increased efficiency, fuel cells, and fossil fuel exploration and production. FE has responsibility for the five storage facilities in the Strategic Petroleum Reserve, the nation's emergency crude oil supply. It also manages three commercial oil fields, and the Naval Petroleum and Oil Shale Reserves. FE's fiscal year 1998 budget is $367 million, 2.2% of DOE's total appropriations. Research activities are performed at four technology centers.

The mission of the **Office of Civilian Radioactive Waste Management (OCRWM)** is to dispose of the nation's commercial and defense spent nuclear fuel and high-level radioactive waste. OCRWM is responsible for the construction and operation of the nation's high-level radioactive waste repository. OCRWM is evaluating the Yucca Mountain site in Nevada to determine whether it is suitable for a geologic repository. Even if the site is determined to be suitable, the repository will not be ready to accept defense waste or commercial spent fuel until sometime after 2010. In the event that an interim storage facility is required, OCRWM will be responsible for constructing and operating such a facility. OCRWM's fiscal year 1998 budget is $346 million, about 2.1% of DOE's total appropriations.

The mission of the **Office of Fissile Materials Disposition (MD)**, in conjunction with the Office of Environmental Management, is to provide for the safe long-term storage of all weapons-usable fissile materials and for the safe disposition (use, dispose of, sell, or otherwise securely eliminate from DOE inventories) of surplus fissile materials. These materials, which exist in several different physical forms (liquid solutions, oxides, metals, and so on), primarily consist of plutonium and highly enriched uranium. MD has developed a different set of strategies for disposition of each material. These disposition strategies will involve the construction of facilities to store, treat, and package fissile materials. While MD's fiscal year 1998 budget is approximately $100 million, less than 1% of DOE's total appropriations, it is expected to increase considerably in future years.

Note: All FY 1998 budget numbers are from U.S. Department of Energy, "FY 1999 Control Table by Organization," in *FY 1999 Budget Request to Congress* (Washington, D.C., February 1998).

The department remains in many ways more an assortment of loosely bound agencies, organizational cultures, and geographic operations than an institution with a unified mission. This institutional legacy raises important questions about the department's ability to address long-term stewardship concerns:

- Is DOE adequately taking into account long-term stewardship considerations in the conduct of its missions?
- Does DOE have the institutional wherewithal to initiate a program of long-term stewardship at its sites?
- Does DOE have the capacity to conduct long-term stewardship at sites, given its multiple and sometimes conflicting missions?

Embedded in today's policy choices are opportunities to decrease the future costs of long-term stewardship. Several major DOE program strategies call for operations at many sites across the country, when concentrating these activities at fewer sites might result in lower stewardship costs over the long term. A few examples follow.

- *Nuclear Stockpile Stewardship and Management.* Current plans of the Office of Defense Programs call for weapons-related activities at eight sites— three national labs, the Nevada Test Site, and four stockpile management sites.[35]

- *Fissile Materials Disposition.* The recently selected strategy for fissile materials disposition (plutonium and highly enriched uranium) will involve nearly a dozen sites and a substantial number of disposition technologies.[36]

- *Waste Management.* DOE's plans for managing wastes generated from past and future nuclear defense and research activities consist primarily of decentralized and regionalized approaches. According to current plans, the Office of Environmental Management will treat, store, and dispose of five types of radioactive and hazardous wastes at dozens of sites across the United States.[37]

EM does have a small ($1.4 million in FY 1998) "stewardship" program, based at its Grand Junction office.[38] This office is currently responsible for long-term surveillance and monitoring (LTSM) of remediated uranium mill tailings sites and a closed commercial low-level radioactive waste disposal site. Stewardship activities for these sites are required by two federal laws, the Uranium Mill Tailings Radiation Control Act of 1978 and the Nuclear Waste Policy Act of 1982.[39] In the near future, Grand Junction expects to receive

---

[35]U.S. Department of Energy, *Final Programmatic Environmental Impact Statement for Stockpile Stewardship and Management* (Washington, D.C., September 1996.)

[36]U.S. Department of Energy, *Storage and Disposition of Weapons Usable Fissile Materials: Final Programmatic Environmental Impact Statement* (Washington, D.C., December 1996).

[37]U.S. Department of Energy, *Final Programmatic Environmental Impact Statement for Managing Treatment, Storage, and Disposal of Radioactive and Hazardous Waste* (Washington, D.C., May 1997).

[38]Albuquerque Operations Office, Grand Junction Office, *Long-Term Surveillance and Maintenance,* Brochure (Grand Junction, Colo.: U.S. Department of Energy, 1997).

[39]*Uranium Mill Tailings Radiation Control Act of 1978,* Public Law 604, 95th Cong., 2nd Sess. (November 8, 1978); and *Nuclear Waste Policy Act of 1982,* Public Law 425, 97th Cong., 1st Sess. (January 7, 1983).

responsibility for at least fifty additional sites, including small waste disposal sites and decommissioned nuclear facilities. For the larger sites, however, there is currently no clearly defined stewardship program, nor a clear assignment about what office—in EM or elsewhere in DOE—is responsible for addressing stewardship issues. The most recent major EM policy document, *Accelerating Cleanup: Paths to Closure*, makes clear that DOE's current long-term stewardship life-cycle cost estimate of under $5 billion is based on limited information provided by many sites and does not include the total costs of stewardship.

A key question is whether DOE can successfully *initiate* a program of long-term stewardship from within. DOE has pursued major new initiatives in the past, but, as in many large bureaucracies, these have generally been in response to external stimuli. DOE initiated a large-scale environmental cleanup program only when the extent of environmental, health, and safety problems across the weapons complex came to public light as the Cold War was ending, and after successful legal action forced DOE to reckon with federal regulators.[40] The Grand Junction LTSM program was created to meet legislative requirements. Absent external forces, such as new legislation, it is questionable whether DOE will launch a comprehensive program of long-term stewardship for the sites in the weapons complex. Most likely, DOE will need to continue to tackle stewardship issues internally until some kind of external action creates the political and bureaucratic will necessary to create a stewardship program.

Recently, the issue of stewardship at DOE sites has begun to get some attention: EM has a staff office funding internal and external research and analysis on this issue (including this report), and the Environmental Management Advisory Board, which is a group of independent experts who advise the EM assistant secretary, recently created a long-term stewardship subcommittee. Still, there is little in the way of a coordinated departmental strategy for long-term stewardship, and little in DOE's history to suggest that it will be able to create a successful program, absent external pressure.

One of the reasons is political. The EM program is under tremendous pressure to show progress, and this means "cleanup." It would be difficult for the department to propose to Congress a new, expensive stewardship program to continue to take care of EM sites, when DOE has been pushing the notion that site cleanup has been achieved, or will be soon.

---

[40]*Legal Environmental Assistance Foundation v Hodel*, 586 F Supp 1163 (ED Tenn D Ct 1984).

Another key question is whether, even if there is a clear stewardship mandate, DOE is likely to be able to successfully *implement* such a program. Myriad reports, both internal and external, suggest that there are many internal barriers within DOE to implementing effective management strategies to address important policy imperatives. As noted earlier, DOE is a hodgepodge of different offices with four quite distinct missions: energy resources, national security, science and technology, and environmental quality. Sometimes these missions come into conflict at individual sites. A tremendous amount of time and energy can go into brokering internal DOE turf battles, rather than moving forward to address the problem at hand. This is the result of both a lack of a clear internal process for brokering disputes and a lack of shared mission among offices. The most obvious example, in the EM context, has been the ongoing dispute between DP and EM over which office would have responsibility, including financial responsibility, for DP facilities that become obsolete in the future. This disagreement continues today.

Several other often-noted organizational factors may also impede DOE efforts to successfully implement long-term stewardship, absent a clear political mandate. These include the duplication of functions across program offices and between headquarters and field offices, enormous political pressures from interest groups and local communities to use the department as a "jobs factory," the applicability of different statutes to different sites, and the extensive use of contractors to carry out DOE missions without adequate federal employees to oversee their work.

The lack of a specific legislative environmental mandate—and the enforcement and public information requirements that go with it—in conjunction with a tangled web of regulations is one of the reasons why it has been so difficult to hold DOE accountable for its actions. Major substantive legislation would provide a vehicle for both internal and external accountability. Instead, DOE is subject to the whims of the annual defense authorization bills, which often include substantive riders, are technically in force for only one year, and change from year to year. While major environmental laws, such as Superfund and RCRA, *apply* to DOE sites and facilities, there is no organic (or integrated) piece of legislation that governs environmental management activities at DOE sites.[41] Thus,

---

[41]Katherine N. Probst, Carolyn A. Pilling, and Karen Turner Dunn, *Cleaning Up the Nuclear Weapons Complex: Exploring New Approaches* (Washington, D.C.: Resources for the Future, July 1996).

each year, DOE is subject to a new set of initiatives through the appropriations process, rather than being governed by an overarching piece of environmental legislation meant to endure for years. In sum, it is clear that the question of what role DOE has in developing and implementing a program of long-term stewardship must be considered within the context of the entire Department of Energy. Ideally, the issue of the need for long-term stewardship should be examined in an even broader context, as many kinds of contaminated sites—both private and federal—are going to need long-term stewardship to ensure protection of public health and the environment.

III. Key Functions of a Long-Term Stewardship Program

The slow but continual action of natural processes and human activity will, in all likelihood, disturb some of the hazards at the sites in the weapons complex after cleanup activities are completed. These disturbances could eventually lead to the introduction of radioactive and hazardous materials into the environment, perhaps in substantial amounts. Thus, for today's activities to continue to be protective of human health and the environment in the long term, certain follow-up activities or stewardship functions must be undertaken.

We envision these activities to include (1) site monitoring and maintenance, (2) application and enforcement of institutional and other controls, (3) information management, and (4) environmental monitoring.[42] Some of these activities are going on now, under existing programs, but they are not part of an integrated stewardship approach. The purpose of this section is to lay out the scope of needed stewardship functions that should apply at all sites in a coherent framework, rather than the current, more fragmented approach.

---

[42]Some would include research and development as a fifth element of long-term stewardship. We have not included research and development as a stewardship function as it is not an integral part of day-to-day stewardship activities.

Site Monitoring and Maintenance

At Department of Energy (DOE) sites that continue to pose current or future risks, in one form or another, a wide range of activities that can be categorized as site monitoring and maintenance will be necessary. These activities will be especially important at those parcels of land that could pose high risks, most notably at some of the major Environmental Management sites, such as Hanford, Savannah River, Oak Ridge, and the Idaho National Engineering and Environmental Laboratory.

Site monitoring refers to periodic inspections to verify that engineered structures and barriers constructed to contain weapons complex hazards and/or to limit human access to those hazards have not been compromised. The many waste disposal facilities, earthen mounds, entombed facilities, and other engineered structures, as well as site infrastructure, fences, warning markers, and other physical constructions will need to be examined and assessed on a periodic basis to determine whether they require maintenance or repair.

Site maintenance activities will also be required at many sites. In addition to the maintenance and repair of structures, such as waste disposal units, that routinely require attention, site maintenance includes maintaining signs, markers, and other systems designed to warn people about remaining site hazards, and ensuring that fences and other barriers are in good repair. Already, markers installed just twenty years ago to mark buried wastes are showing signs of disrepair.[43]

These types of activities will be required as long as materials, wastes, contamination, or facilities at a site pose a potential current or future risk. Long-term site monitoring and maintenance activities will not only help to postpone the deterioration of closed waste disposal facilities, engineered barriers, and site infrastructure; these activities will also help ward off inadvertent and intentional human activities that could result in human or environmental exposure to remaining site hazards.

Institutional and Other Controls

A necessary and important component of a long-term stewardship program will be the set of institutional and other controls implemented at DOE sites to prevent or limit the exposure of human beings to post-cleanup hazards. Insti-

---

[43]Office of Environmental Management, *Closing the Circle*, p. 100.

tutional controls can be defined as legal and informational mechanisms that are employed to limit human activities on or near a given parcel of land.[44] At a contaminated site, institutional controls can be used to prevent or limit human exposure to site contamination by prohibiting particular uses of the site.[45] For example, at a parcel of land with contaminated groundwater, appropriate institutional controls would be those required to prevent drinking-water wells from being drilled. For areas with contaminated soil, institutional controls proscribing excavation at the site could be implemented to prevent human exposure.

Institutional controls can generally be categorized as one of two types: *local government* controls or *proprietary* controls. Local government controls include zoning restrictions, permit programs, well-drilling restrictions, and other restrictions that are traditionally established under the authority of local governments. Proprietary controls are legal devices, such as deed restrictions, easements, and restrictive covenants, that are based on state property law and are used to restrict the private use of property. All of these mechanisms have as their goal restricting the use of a given parcel of land—but the method by which that goal is pursued and who has authority to enforce the restrictions vary depending on the particular control implemented.

Historically, it has been difficult to ensure that institutional controls remain effective over time.[46] In the case of government controls, local political processes through which land-use controls are established are unpredictable. Zoning restrictions, for example, typically change over time, as in some sense, they were intended to. This makes zoning a less than satisfactory form of institutional control, at least on its own.

Some proprietary controls are more enduring than others. For example, most easements, which are deed restrictions that enable their holders to restrict activities on a property, "run with the land." That is, they are binding on the new

---

[44]George Wyeth, "Land Use and Cleanups: Beyond the Rhetoric," *Environmental Law Reporter* (July 1996): 10362.

[45]Robert Hersh, Katherine Probst, Kris Wernstedt, and Jan Mazurek, *Linking Land Use and Superfund Cleanups: Uncharted Territory* (Washington, D.C.: Resources for the Future, June 1997).

[46]Discussion of this problem can found in Wyeth, "Land Use and Cleanups"; Hersh and others, *Linking Land Use;* and John Pendergrass, "Use of Institutional Controls as Part of a Superfund Remedy: Lessons from Other Programs," *Environmental Law Reporter* (March 1996); as well as a number of reports on Yucca Mountain issued by the National Research Council and others.

titleholder when title to a property is transferred. However, some proprietary controls, such as restrictive covenants, do not "run with the land," and thus are not binding on the new titleholder. The legal basis of proprietary controls can vary from state to state. Of course, sometimes, even when future use of land is clearly restricted, these restrictions are ignored. The most famous example of this violation of institutional controls is Love Canal.

For successful implementation of institutional controls under a long-term stewardship program, these and other problems—such as difficulties keeping track of property ownership, enforcement of the controls, and state-to-state variations in property law—would need to be resolved. It is critical that mechanisms be implemented to ensure that any renters or purchasers of DOE land and facilities be informed of hazards at sites they lease or own. Another, and perhaps more effective, form of institutional control available for federal facilities is continued federal government ownership and control. The federal government can restrict the use of land, surface water, and groundwater on land it owns and controls (although there are some very difficult issues for tribal lands). Federal land ownership as a form of institutional control would also have the benefit of placing long-term financial responsibility for ensuring controls on the federal government. However, even continued federal ownership affords no perfect guarantee. The U.S. government could decide to subsequently sell or lease the land in question, and may or may not enforce needed controls over the time period required.

Other types of controls employed for long-term stewardship purposes could include physical objects, structures, or mechanisms that deter intrusion into areas, both above- and below-ground. These controls could range from simple signs indicating the presence of hazards to engraved stone monoliths to recorded warning messages. In the context of enduring weapons complex hazards, these types of controls intended to function well into the future are often referred to as "markers" or as "passive" institutional controls.[47]

Information Management Systems

Information management systems will be necessary to store, preserve, and inte-

---

[47]Kathleen M. Trauth, Robert V. Guzowski, Chris G. Pflum, and Ronald J. Rodriguez, *Effectiveness of Passive Institutional Controls in Reducing Inadvertent Human Intrusion into the Waste Isolation Pilot Plant for Use in Performance Assessments*, WIPP/CAO-96-3168 (Carlsbad, N.M.: U.S. Department of Energy, November 14, 1996).

grate information about a wide range of issues associated with remaining site hazards. Information management systems for long-term stewardship must be capable of efficiently acquiring and integrating new information, while storing and preserving previously acquired information.

At a site-specific level, basic information should be retained and preserved about several aspects of each hazard, including its physical and chemical characteristics; how it was created and whether it has changed over time; and the measures, both engineering and institutional, that have historically been applied to contain it. At a more general level, information management systems might be expected to retain and preserve crosscutting information that concerns broad issues, such as demographic and economic change in the vicinity of hazards, regional environmental and health monitoring data, the success or failure of hazard management strategies, and general institutional history and lessons learned.

Information management systems must manage stored information in such a manner that present and future generations will have ready access to important information. The availability and accessibility of information to the public will be crucial for program success. The challenge of maintaining operational and effective information management systems for such complex information over long time periods is without precedent.

Technological change, leading to the obsolescence and replacement of information management technologies, may result in the unintentional loss of information, especially if high-technology systems are relied on exclusively. Low-technology approaches—such as handwritten records and files—have their own difficulties, including inefficiencies, limited information accessibility, and the absence of an enduring medium on which information can be reliably stored. Most likely, a hybrid information management system involving both high and low technologies is needed. Most important is that one or more institutions be designated responsible for developing such a system now.

In addition to technical challenges posed by long-term information management, economic, social, political, technological, and cultural changes and disruptions that are not currently foreseeable could complicate the implementation of information management activities. Wars, civil unrest, economic depression or collapse, changing political systems, growing public aversion to risk, and an evolving English language are all factors that might influence, in one respect or another, long-term information management governing remaining weapons complex hazards.

Ideally, information management systems established as part of a long-term stewardship program should be organic systems that can accommodate the addition of relevant new information, that can adapt over time according to external circumstances, that can compensate for failure of the media on which information is stored, and that can be easily accessed by future generations. Appropriately managed and preserved, information captured by a long-term stewardship program will assist future generations in understanding the nature of the hazards created by nuclear weapons production, the stewardship solutions that were implemented to address remaining hazards, and how to revisit both cleanup and stewardship solutions.

Environmental Monitoring

Environmental monitoring at DOE sites is needed to ensure that information is available to perform risk evaluations over time. By monitoring the environment in the vicinity of weapons complex sites, current and future decision makers could receive early warning of detrimental environmental impacts resulting from remedy and stewardship failures. With that information, they would be in a position to address the negative impacts appropriately and act to limit the future incidence of such effects. Similarly, by tracking the evolution and migration of weapons complex hazards, scientists could inform decision makers of the failure of engineered barriers and migration of contaminants, warn of new threats to human health and the environment, and suggest appropriate measures to counter such threats.

Environmental monitoring at and around DOE sites could provide important information about incremental environmental changes that may be the result of remaining weapons complex hazards. For example, it might make sense to conduct periodic monitoring of contaminant levels in plant and animal organisms, in surface soil, in groundwater, and in surface water. Similarly, monitoring of plant and animal population characteristics might be carried out to detect any changes that might be the result of increased background levels of contamination.

Detrimental environmental effects resulting from weapons complex hazards would be of concern not only because of their impact on environmental quality but also because of their impact on human health. Increased levels of contamination and radioactivity in surface water near DOE sites could be harmful to ecosystems and animals, and to people who rely on those ecosystems and animals for food and habitat. This is a major concern to tribal nations. Since radioactive and hazardous contaminants often accumulate in animals, people feeding on these animals could ingest contaminants through the food chain. Similarly, crops grown in contaminated soil or irrigated with contaminated groundwater might absorb certain radioactive contaminants and metals that might later be ingested by humans.

Among weapons complex hazards, groundwater contamination, which will persist at many DOE sites since it is almost impossible to eliminate with current technology, is particularly difficult to contain and most likely to migrate over time. Therefore, long-term tracking of contaminant plumes will be important to determine the extent to which groundwater aquifers become contaminated and to assess the potential risks posed by such contamination to humans and ecosystems.

IV. Creating a Successful Stewardship Program

Lessons for stewardship of Department of Energy (DOE) sites, especially about the capacity of institutions to endure over time, can be drawn from an interesting analogy—that of cemetery preservation and maintenance. Perhaps the most important similarity between the administration of cemeteries and the stewardship of DOE sites is that the functions associated with both endeavors will need to be implemented far into the future. Some of the functions required for cemetery preservation—cemetery maintenance and monitoring, passive institutional controls (fences, tombstones, access restrictions), and burial records—mirror those needed for long-term stewardship at DOE sites—site maintenance and monitoring, institutional controls, and information management. Unlike DOE sites, however, cemeteries have been a phenomenon in human civilization for thousands of years, and there is a considerable record of experience administering them. Examining the history of cemeteries raises some interesting issues that are relevant to long-term stewardship at the sites in the weapons complex.

The historical record tells us that the fate of these sites is diverse, but rarely what was envisioned at the time they were established. Over time, cemeteries are often vandalized, overgrown with vegetation, built over, moved to another location, or simply abandoned and forgotten. Even some of the most well-known cemeteries experience vandalism or neglect. Examples include the Pere-Lachaise cemetery in Paris and the Congressional cemetery in Washington, D.C. Many of the cemeteries that are well maintained are treated as responsibilities of government agencies or religious institutions. Arlington Cemetery, for example, is maintained by the U.S. Army and has an annual budget of $11 million a year.[48] Many state legislatures require licensed cemeteries to establish a perpetual care fund to prevent cemeteries from falling into a state of neglect.

Although the cemetery analogy has flaws, it demonstrates the difficulty of maintaining an effective institutional presence and program at a given site or set of sites for long periods of time. If, in little more than decades, cemeteries tend to fall into disrepair, what can be expected of DOE sites where long-lived radioactive and other hazards will pose risks to human health and the environment for hundreds if not thousands of years? The cemetery analogy suggests that the major challenge to ensuring a successful long-term stewardship program will be to develop institutions that can endure over time and a community (however defined) that has an interest in ensuring that stewardship responsibilities are not neglected.

It seems obvious that there *will* be a need for long-term stewardship at the sites in the nuclear weapons complex. On this point, there is likely to be little disagreement. Exactly what "stewardship" means and how it should be defined, however, will require some debate. One key issue is whether a stewardship program should be created for all contaminated sites—whether public or private, all federal facilities subject to Superfund, or only for DOE sites. The decision about the scope of a stewardship program has important implications for what organization (or organizations) should implement stewardship functions, and for how a stewardship program should be created and funded.

What is likely to be controversial is the decision of which institution (or institutions) should carry out stewardship responsibilities. Most daunting will be to ensure that the institution charged with stewardship responsibilities has the bureaucratic, political, and financial wherewithal to successfully implement these functions.

Defining the mission of a long-term stewardship program is simple: To take those actions necessary to ensure protection of human health and the environment from hazards that remain at sites in the weapons complex after

---

[48]*Departments of Veterans Affairs and Housing and Urban Development, and Independent Agencies Appropriations Act, 1998,* Public Law 65, 105th Cong., 1st Sess. (October 27, 1997).

"cleanup" (engineering solutions) has been implemented. Actually *creating* an enduring stewardship program is a much more difficult task. We do not make what could only be called a foolhardy attempt to argue for the creation of a permanent stewardship institution. Instead, we take a different tack. While it is critical that a stewardship institution endure in the short run (a lofty enough challenge), it is the longevity of stewardship functions that must be ensured, not the longevity of the institution itself. Thus, as part of the institution's mission, it must continually reevaluate how stewardship functions are being implemented, with the goal of ensuring that its mission will endure until the next generation. A stewardship institution must also have as part of its mandate a commitment to a culture of openness for this goal to be realized.

Four critical steps are necessary to create a successful stewardship program. First, a case must be made for why such a program is necessary. We hope that this paper, along with many others, most notably *Moving from Cleanup to Stewardship*, a draft report put together by DOE's Office of Strategic Planning and Analysis, makes that case.[49] Second, a clear stewardship mandate needs to be established. Third, a specific organization (or part of an organization) needs to be given the responsibility, the resources, and the authority to implement stewardship functions. Fourth, mechanisms must be created to ensure that those organizations charged with implementing stewardship functions are held accountable. In the remainder of this section, we outline options to be considered in each of the steps just identified.

Creating a Stewardship Mandate

To create an enduring stewardship mandate, federal legislation will be necessary. While it is true that DOE could create its own stewardship program, DOE's internal policies and priorities change frequently. Without a legislative mandate, whatever mission and organizational structure is created for stewardship, the

---

[49]Other reports that document the nature of the Cold War's environmental legacy include Office of Environmental Management, *Linking Legacies;* Office of Environmental Management, *The 1996 Baseline Environmental Management Report;* Office of Environmental Management, *Taking Stock: A Look at the Opportunities and Challenges Posed by Inventories from the Cold War Era* (Washington, D.C.: U.S. Department of Energy, January 1996); and Marc Fioravanti and Arjun Makhijani, Ph.D., *Containing the Cold War Mess: Restructuring the Environmental Management of the U.S. Nuclear Weapons Complex* (Takoma Park, Md: Institute for Energy and Environmental Research, October 1997).

program is unlikely to endure or to be successful. Congress will need to enact federal legislation outlining the stewardship mission, identifying the scope of sites to be addressed (such as DOE sites, all federal facilities, all contaminated sites subject to the Resource Conservation and Recovery Act [RCRA] and Superfund, and so forth), setting out the roles and responsibilities of the organization(s) charged with implementing stewardship activities, and creating mechanisms to ensure external accountability. Such legislation would also need to make clear the role of states, local governments, tribal nations, and others as meaningful partners in this endeavor; require an open decision-making process; and, finally, provide for citizen suits to enforce the provisions of the legislation.

A legislative stewardship mandate is needed for four reasons. First, absent legislation, it is unlikely that a coherent and enduring stewardship program will be created. Legislation often provides senior management of federal agencies with a much-needed basis for making major internal management and organizational changes. Clearly, these kinds of changes will be needed for DOE to successfully implement stewardship activities at its sites.

Second, substantive legislation is needed to ensure that a stewardship mandate (and program) withstand the vagaries of changing politics and administrations. While, for example, the Clinton administration might take steps to implement a stewardship program, purely administrative changes can easily be undone by the next presidential administration that follows. The long-term nature of the stewardship challenge requires a major commitment on the part of the federal government. At least in theory, a stewardship institution may need to endure for centuries. History tells us, however, that institutions, even countries, don't last forever. Even cultures, handed down from generation to generation, undergo change. Certainly, there have been dramatic technological, political, and cultural changes in our lifetime. While requiring a stewardship program in federal legislation does not guarantee its survival, it is the best mechanism for creating an enduring federal stewardship mandate.

A third reason for enacting legislation establishing a stewardship mandate is to provide a clear and enduring basis for federal funding. Again, federal legislation does not guarantee federal funding (there are numerous programs outlined in federal legislation that Congress has neglected to fund), but it does place the issue squarely on the appropriations agenda. Garnering the necessary resources to implement a successful stewardship program will be a major challenge. By definition, a long-term stewardship program has as its goal prevention of contamination and exposure in the distant future. As has been seen in the debate about global climate change, it is much more difficult to obtain the necessary resources for these types of future activities than for today's crisis.

As part of developing a stewardship program, it will be important to examine alternative mechanisms for ensuring long-term funding. One mechanism would be to create a federal trust fund, as has been done for Superfund. Of course, the Superfund example (and others) shows that putting aside monies in a federal trust fund does not guarantee they will be used for their intended purpose. The Superfund Trust Fund, for example, has often been used to balance the budget and reduce the need for general revenues, rather than being appropriated for the Superfund program. For private facilities, one common mechanism is to charge a tax or fee on operating facilities to finance a post-closure fund. This option does not exist for federal facilities, as the federal government does not tax itself. In addition, many of the DOE facilities are no longer in operation, and thus there are no "operations" to tax. Requiring a stewardship program in federal legislation is a necessary, although not sufficient, step to increase the chances that such a program will receive adequate funding.

Fourth, and finally, enacting stewardship legislation would provide a clear basis for external accountability. As we have seen in the Office of Environmental Management's brief history, it is terribly difficult for organizations to embrace new missions. The bureaucratic tendency is one of inertia—to keep doing what you've been doing. Clearly articulating the scope and purpose of a stewardship program in federal legislation would provide the basis for congressional oversight. Legislative provisions for citizen suits would ensure public scrutiny.

Designating a Stewardship Implementor

It is important that responsibility for implementing stewardship activities be clearly assigned to a specific organization (or organizations), along with the required resources. In theory, a long-term stewardship program for DOE sites could be implemented under several alternative institutional frameworks. A single institution could be charged with implementing stewardship activities, or many institutions could be involved. Stewardship functions could be implemented solely by the federal government or by a combination of federal, state, and local agencies. Existing government agencies could be charged with stewardship responsibilities or new stewardship organizations could be created. From a pragmatic standpoint, however, the chances of garnering the political and bureaucratic support to create a stewardship mandate are probably best if stewardship functions are assigned to an existing organization. The political and bureaucratic obstacles to creating a new federal agency are immense.

Due to the complexity, vast size, and geographic distribution of the weapons complex; the national security risks present at many sites; and the long-lived nature of the radioactive hazards present on DOE sites, the federal government must have primary responsibility for long-term stewardship. Different DOE sites will clearly have different stewardship needs. Some sites might warrant a federally implemented stewardship program, while others might be better suited to a state or local government steward. Most would agree that a site as large and complex as Hanford needs some federal presence, in partnership with the state environmental agency. Smaller sites may not demand a federal role. Differences among sites should be taken into account in developing a stewardship program. It may well be that state, tribal, and local governments can—and should—take on stewardship responsibilities at certain sites. This important issue is not addressed in this paper, but should be the topic of additional research and discussion as part of the process of developing a long-term stewardship program.

The primary locus for stewardship should be the federal government, as one of the most enduring of today's institutions, other than religious entities. State agencies also should be involved in stewardship and should take the lead in oversight of stewardship activities as occurs under RCRA. However, federal agencies should (at least initially) have the primary responsibility for stewardship at DOE sites, with the stewardship mission, goals, and objectives set out in federal authorizing legislation. As important, federal appropriations specifically earmarked for stewardship activities will be needed, both to fund the program and to confirm the federal government's commitment to long-term stewardship. Some federal stewardship role is necessary to ensure that a coherent national program is created, and to garner needed federal resources.

The next major question is whether DOE should implement stewardship functions at its own sites, or whether some other federal agency might be more appropriate. In thinking about what agency should be given this responsibility it is important to keep in mind two criteria: trust and expertise.

- *Trust.* Trust refers to the public's confidence that the institution charged with stewardship responsibilities is, in fact, committed to that mission. It also relates to public trust in the stewardship institution to carry out its responsibilities in an open and fair manner. In the stewardship context, one measure of credibility is likely to be the reality—and perception—that the stewardship institution has the independence needed to take actions it deems necessary to ensure protection of public health and the environment.

- *Expertise.* The stewardship functions described in Section III—site monitoring and maintenance, implementation and enforcement of institutional controls, information management, and environmental monitoring —require a variety of different types of expertise. The institution charged with stewardship functions needs to have the necessary technical and legal expertise to carry out these responsibilities.

There are several logical candidates among existing federal agencies for the federal stewardship "implementor":

- U.S. Department of Energy's (DOE) Office of Environmental Management;
- U.S. Environmental Protection Agency's (EPA) Office of Emergency and Remedial Response;
- U.S. Department of Defense's (DOD) Army Corps of Engineers; and
- U.S. Department of the Interior's (DOI) Bureau of Land Management.

We briefly describe the pros and cons of each. In addition, we explore the option of creating a new office within one of these existing agencies to implement stewardship functions at DOE sites. There are, of course, other federal agencies (and other offices in the agencies examined) not included here, such as the National Park Service, that could also be effective stewardship implementors.

DOE's Office of Environmental Management (EM)

At first glance, the most qualified organization to carry out long-term stewardship would appear to be EM. This is because EM has had the primary responsibility for a wide range of environmental management activities at DOE sites— including some aspects of nuclear materials management—since it was created in 1989. In addition, EM has a small stewardship program at its Grand Junction office. Thus, the office has hands-on experience, as well as much of the technical and policy expertise needed to implement stewardship activities. EM staff are already located at all the major sites in the weapons complex and, for many sites, EM provides the major budget authority.

DOE's environmental management program has been the subject of a tremendous amount of criticism over the years. The lack of public trust in DOE has been noted in myriad reports and articles about the department and its pre-decessors. [50] Many of those living near DOE sites might well question whether DOE *does* have the staff expertise and culture to successfully implement a stew-ardship program. The endless saga of missteps at some of the sites and the fail-ure to disclose key information to the public have led many to question whether DOE is the right agency to manage site cleanups, much less take on stewardship responsibilities. There are repeated instances of DOE operations offices appear-ing reluctant to embrace an environmental management mission, or not taking needed steps to improve their programs.[51]

In addition, Congress has created obstacles in the annual appropriations bills to putting people with the right training in charge of environmental man-agement programs.[52] The National Defense Authorization Act of 1993 requires, for example, that DOE consult with the Department of Labor, state and local governments, and community groups if it believes a restructuring of the work-force is needed at defense facilities. In addition, the law requires that employees whose employment is terminated receive preference in any hiring that is done. This has the effect of giving employees who are well qualified for the tasks they used to do at DOE facilities hiring preferences for DOE's new mission at sites—environmental management—which they often have little training or experience to perform. The fact that DOE will retain ongoing missions at many sites and that there are national security issues relating to nuclear materials makes some DOE role almost unavoidable. If DOE is given such a role, however, making sure that this role is successfully carried out will almost certainly require an internal restructuring of DOE or the creation of a new office within the department.

EPA's Office of Emergency and Remedial Response

Another obvious contender for the lead implementing agency is EPA's Superfund office, formally known as the Office of Emergency and Remedial Response. EPA implements cleanups at almost one-third of the sites on the Superfund National

---

[50]For example, see Secretary of Energy Advisory Board, Task Force on Radioactive Waste Management, *Earning Public Trust and Confidence: Requisites for Managing Radioactive Wastes* (Washington, D.C.: U.S. Department of Energy, November 1993).

[51]See, for example, U.S. Department of Energy, *Review of the Federal Management of the Tank Waste Remediation System (TWRS) Project* (Washington, D.C., January 15, 1998).

[52]*National Defense Authorization Act for Fiscal Year 1993,* Public Law 484, 102nd Cong., 2nd Sess. (October 23, 1992), § 3161.

Priorities List (NPL) and oversees site cleanups at many more.[53] As the lead federal agency for Superfund, EPA staff have the technical, legal, and policy expertise needed to implement a site cleanup program. And, EPA has good working relationships with many state agencies, as a result of state implementation of RCRA. In addition, EPA has two offices (the Federal Facilities Restoration and Reuse Office and the Federal Facilities Enforcement Office) whose staff are knowledgeable about many of the former weapons production sites, as all the major DOE sites are on the NPL. Finally, because of EPA's long-term responsibilities for some NPL sites (these are referred to as "fund-lead" or "orphan sites"), it will eventually need to develop its own long-term stewardship program for contaminated sites, although it has not yet done so.

There are some reasons, however, to be concerned about EPA as the stewardship implementor at DOE sites. First and foremost is that EPA's primary function is that of a "regulator" rather than an implementor. Giving EPA implementation responsibility for stewardship would lead in some ways to self-regulation, as EPA would be the regulator under Superfund and RCRA of many site activities. In fact, some would argue that this tension between the Superfund program and the RCRA program has already led to confusing policy regarding contaminated site cleanup. Second, EPA does not have a very large budget compared with other federal agencies, which raises the question of whether it would be given adequate resources to implement a stewardship program. Third, the Superfund program—like EM—has received much criticism for its implementation of its cleanup responsibilities. Finally, there is tremendous political pressure in Congress to sharply curtail the Superfund program, as evidenced by efforts to limit new additions to the NPL. It is also politically unpopular to talk about having a Superfund program "forever," as would be required for stewardship. However, arguably there will need to be some kind of long-term Superfund program, at the state or federal level, to make sure that institutional controls are complied with over the coming decades.

DOD's Army Corps of Engineers (the Corps)

The Corps is best known for its large-scale construction and engineering projects. The Corps' Environmental Division manages over $1 billion annually in

[53]U.S. General Accounting Office, *High Risk Series: Superfund Program Management,* GAO/HR-97-14 (Washington, D.C., February 1997).

environmental cleanups. These include over 8,700 formerly used defense sites owned or operated by various Department of Defense (DOD) agencies. The Corps also conducts cleanups for civilian agencies, including EPA. There is some precedent for the Corps' taking over responsibility from DOE for environmental management activities, as the Formerly Utilized Site Remedial Action Program was recently transferred from EM to the Corps. As a result, the Corps is likely to develop more expertise over the coming years in dealing with radioactive contamination, at least at small sites.

Assigning stewardship implementation to the Corps would have several drawbacks, however. Despite the Corps' experience with radioactive and hazardous cleanups, it has not dealt with sites that present the range of complexities that exist at DOE sites. Moreover, although DOD, of which the Corps is a part, manages large installations and other properties, land stewardship is not one of the department's primary missions. In fact, under the Base Closure and Realignment Commission (referred to as "BRAC") program, DOD is actively trying to get out of the land management business, unless there are active DOD missions at the sites.

DOI's Bureau of Land Management (BLM)

Land and resource stewardship is at the heart of the Bureau of Land Management's mission. BLM currently manages 270 million acres of public land on which multiple commercial and recreational activities take place. The bureau also maintains millions of pages of historic land documents that have been acquired by BLM or its predecessors and is currently struggling with the development of an advanced system to manage these records. However, BLM has little experience in dealing with hazardous and radioactive materials, wastes, or contamination. In addition, if one of the jobs of site stewardship is to protect the land, BLM may face some credibility problems as a stewardship agency as it has been criticized for poor management of its lands. BLM's role, as set out in legislation, includes releasing (and, in fact, encouraging the release of) lands for mineral exploration and other resource utilization purposes.

New Office within DOE, EPA, DOD, or DOI

A new long-term stewardship office could be established and situated in one of the aforementioned agencies. The creation of a new office charged specifically with stewardship responsibilities would have the advantage of avoiding some of the drawbacks of existing offices discussed above. Such an approach would also provide a central focus, and visibility, to stewardship activities. And, at least in theory, developing a stewardship ethic and an appropriate organizational culture would be less difficult for a newly established stewardship office. However, as the history of EM has shown, simply creating a new box on the organization chart is not enough. Care must be taken in creating a new office (in whatever agency) to make sure that it really does "march to a different drummer." A new office would, of course, be successful only if it were given a clear and unambiguous stewardship mandate, strong leadership, and the necessary staff and other resources to carry out its mission.

Conclusion

As the foregoing discussion shows, there is no obvious "best" candidate for implementing stewardship activities at DOE sites. The following table summarizes our evaluation of the different federal agencies according to the two criteria articulated earlier: trust and expertise.

	DOE	EPA	DOD	DOI
Trust	Low	Medium	Medium	Low
Expertise	High	Medium	Low	Low

Certainly, DOE will continue to implement stewardship functions in the short term. Steps can and should be taken to improve the credibility of DOE's activities in this area. Beyond that, the question of what agency (or agencies) should implement stewardship functions may well rest with the decision of whether now is the time to initiate a stewardship program for DOE sites only, for federal facilities more generally, or for all contaminated sites, public and private, that require some sort of post-closure care. If the scope of a stewardship program is broadened beyond DOE sites, then the argument becomes stronger for shifting implementation responsibility to another agency, such as EPA. Most likely, DOE will be the lead stewardship agency for its sites, at least in the near future. If this is indeed the case, though, steps need to be taken to ensure that DOE is held accountable for its actions.

Ensuring External Accountability

The final component of a successful stewardship program is an effective mechanism for ensuring external accountability. As noted earlier, creating a legislative mandate is one of the best ways to accomplish this goal. This legislation should include provisions requiring periodic reports to Congress on the progress of the stewardship program, public involvement in key stewardship decisions, and citizen suits if government agencies charged with stewardship responsibilities do not comply with the legislative requirements.

While some federal role in stewardship will be necessary, states, localities, tribal nations, and the general public must be meaningfully involved in the development, implementation, and oversight of stewardship activities. The involvement of stakeholders in stewardship will help increase public trust in a stewardship program and ensure much-needed external accountability. History suggests that the involvement of these other entities is critical to keeping the federal system "honest."

The creation of a formal mechanism for independent oversight of stewardship activities also is needed to assure external accountability. There are two basic models for independent oversight: the regulatory model and the independent oversight model.

Under the regulatory model, typified by EPA and the Nuclear Regulatory Commission (NRC), the regulator sets standards in regulations and then has enforcement powers to hold those subject to the regulations responsible for complying with these standards. DOE, for example, must comply with hazardous waste management regulations promulgated by EPA to implement RCRA, as well as with EPA requirements for cleanup of contaminated sites under the Comprehensive Environmental Response, Compensation, and Liability Act (CERCLA).

The two obvious candidates for providing some kind of external oversight of stewardship implementation are EPA and the NRC. While each agency would bring important regulatory expertise to the job, neither is a perfect fit. If EPA were to be given the job of overseeing stewardship, it would need to enhance greatly its expertise regarding nuclear materials and wastes. If NRC were given this role, it would need to expand its knowledge of the more traditional environmental concerns.

Another model for stewardship oversight would be to create a small independent agency. Such an agency could be modeled on the Defense Nuclear Facilities Safety Board (DNFSB), an independent agency charged with overseeing DOE's nuclear safety activities. An independent oversight agency tends to operate on a more ad hoc basis. It typically has a broad charter and can review activities of the agency subject to its oversight. It can make recommendations for changes that are needed, but typically has little enforcement authority, other than embarrassment and moral suasion.

An independent stewardship oversight agency would have as its sole mission oversight of stewardship activities (at DOE sites, or a broader set of sites), a board of independent experts, and a small staff (DNFSB's staff, for example, is limited to 150 people), with the specific technical, environmental, and legal expertise to do the job. Such an agency could be created at the federal level, or a joint federal-state oversight commission could be created.

Clearly, the states, too, must have a role in stewardship oversight. One of the strengths of the current environmental regulatory structure vis-à-vis DOE is that there is dual regulation, with the states often in the lead under RCRA and EPA typically in the lead implementing Superfund. This has been crucial to effective oversight of DOE activities. In addition, formal mechanisms would be needed to enable some stewardship oversight by tribal nations and localities, if appropriate. The whole question of how tribal nations might help implement stewardship is a very interesting issue. These societies have cultures that are much older than the federal government and could provide an important component of long-term stewardship.

In the following paragraphs, we sketch out the basic pros and cons of the three major institutional alternatives for stewardship oversight: EPA, NRC, or a new independent oversight agency.

Environmental Protection Agency

EPA is the only federal agency whose sole mission is protecting human health and the environment. EPA already has experience with environmental regulation at weapons complex sites as the lead federal regulator for environmental activities conducted under Superfund and RCRA. Thus, agency staff already have expertise regarding many of the problems and challenges facing the sites in the weapons complex. In addition, federal policy about the role of future land use at contaminated sites and the use of institutional controls is likely to be driven by the Superfund program.

On the downside, many view EPA as a sometimes weak regulator of other federal agencies. While this could in theory be fixed in legislation, DOE is in many ways a much more powerful agency than EPA. DOE's budget is three times larger than EPA's, and DOE is a cabinet-level agency, which EPA is not. EPA has limited statutory authority over radioactive waste and materials, which are present at most DOE sites. Finally, EPA has a large portfolio of environmental regulation yet operates on a tight budget.

Nuclear Regulatory Commission

With statutory authority over the commercial nuclear industry, NRC has acquired in-depth knowledge of, and experience with, nuclear materials, radioactive wastes, and nuclear facilities from the commercial nuclear industry. DOE recently recommended that NRC be given responsibility for external regulation of DOE nuclear safety at DOE facilities.[54] NRC has regulatory authority over closed commercial low-level waste disposal sites, remediated uranium mill tailings sites, and decommissioned commercial nuclear facilities. When hazards remain after remediation or decommissioning is complete, these facilities must comply with NRC requirements for long-term custody and surveillance. NRC, along with EPA, has regulatory authority over the high-level waste repository. In addition, NRC works with "agreement states" to implement NRC programs.

However, NRC has limited experience with issues related to the sites in the weapons complex, or with hazardous wastes and hazardous waste management facilities. And a recent policy tussle between EPA and NRC over radioactive cleanup standards raises serious questions about whether NRC would have credibility as an overseer of stewardship activities. NRC has had credibility problems over the years as a regulator of commercial nuclear facilities and is unlikely to be perceived as a strong advocate of compliance with the nation's environmental laws, even though it has stepped up its enforcement activities and improved its credibility in recent years. Finally, with a relatively small annual budget of less than a half billion dollars raised almost entirely by user fees, finding a stable source of additional funding for oversight of DOE stewardship activities could prove difficult.[55]

Independent Stewardship Agency

Regulatory agencies tend to create unwieldy bureaucracies to achieve their goals. The very structure of a regulatory agency depends on writing regulations generally applicable to a number of types of facilities. This makes it difficult to take quick action and create approaches tailored to the situation at hand.

---

[54]U.S. Department of Energy, *Report of Department of Energy Working Group on External Regulation*, DOE/US-0001 (Washington, D.C., December 1996).

[55]U.S. Nuclear Regulatory Commission, *Budget Estimate, Fiscal Year 1999*, NUREG-1100, Vol. 14 (Washington, D.C., February 1998).

Creating an independent stewardship agency, coupled with a strong enforcement mechanism, would allow the structure and staffing of the organization to be tailored to the specific problems at DOE sites (or at contaminated sites more generally). In addition, it might be possible to create a joint federal-state commission to conduct stewardship oversight, which would have the advantage of giving the states an integral role in stewardship from the beginning. On the downside, Congress is unlikely to want to create a new agency or commission, and existing agencies might see a new organization as stepping on their turf.

Conclusion

The charge of ensuring external oversight of stewardship will necessitate, in one form or another, a new organization—either within an existing agency or as a new agency. Even if the scope of a stewardship program is confined to DOE sites, there is currently no organization that has the full complement of skills needed to conduct an effective stewardship oversight program. No matter which organization is given responsibility for overseeing stewardship implementation, it would need to acquire some additional expertise to successfully oversee long-term stewardship activities at DOE sites. Right now, EPA is much more actively involved in overseeing EM sites than NRC because of its role in implementing CERCLA and RCRA, as well as the other major environmental laws. In addition, EPA has more credibility than NRC as an environmental regulator. Although NRC has been involved in activities related to nuclear wastes, it has little experience with EM sites.

For these reasons, EPA is probably a more likely agency to conduct external oversight than NRC. Experience, however, has shown that EPA is a more effective regulator of DOE when states also have regulatory authority, as is true under RCRA. This suggests that stewardship oversight might be best served by some sort of joint oversight, conducted by both the federal and state government. An independent agency or commission, however, is still an attractive option as it could be created specifically to address stewardship needs. The downside of such an approach is that a new commission is more likely to be eliminated with a political wave of the hand than a major federal agency.

V. Recommendations for Next Steps

Addressing the issue of long-term stewardship is critical to ensuring the success of the Department of Energy's (DOE) environmental management program. There is likely to be little disagreement among experts that there is a need for the types of activities described in this report. Still, agreement about the need for a stewardship program does not automatically translate into the necessary political and bureaucratic actions to make that program a reality. Money is likely to be a critical stumbling block. While tomorrow's stewardship costs may be much smaller than today's cleanup costs, they are a long-term responsibility and liability of the federal government. In many ways, a frank disclosure of future stewardship costs flies in the face of recent Environmental Management (EM) initiatives that have been aimed at showing Congress (and others) that DOE is making progress in cleaning up its sites, and that there is an end in sight to the EM program and federal appropriations on the order of $6 billion a year. Thus, it may be a difficult time for DOE to acknowledge that stewardship at DOE sites is a long-term federal responsibility, and that federal appropriations will be necessary for an effective and enduring stewardship program.

DOE is already implementing stewardship activities at some of the smaller sites in the weapons complex. DOE needs to take action now to show leadership on this issue and to start to put in place information and strategies that provide a sound base for developing a comprehensive stewardship program for all DOE sites. All that being said, however, serious consideration should be given to the question of whether DOE should have long-term responsibility for stewardship, given its past record. Clearly, DOE currently has responsibility for stewardship (even if not explicitly recognized for all sites), and it will have a major role implementing these responsibilities in the short term. A key organizational question is whether DOE should continue to have a major role in stewardship of its sites, or whether responsibility for long-term stewardship should be transferred to another federal agency, or to state agencies, for certain sites. This question deserves to be the subject of a thorough debate among all the stakeholders at DOE sites. Most likely, DOE will continue to have responsibility for implementing stewardship activities at its sites. Increased external oversight—by the U.S. Environmental Protection Agency (EPA), states, or some other agency—will be needed to hold DOE accountable and increase public confidence that important post-closure activities are, in fact, being conducted.

In some sense, the most important "next step" is to stimulate a public dialogue about the need to address long-term stewardship at the nuclear weapons complex sites and about the scope of coverage of a stewardship program. Addressing these issues is the responsibility of all stakeholders—DOE, EPA, state agencies, local governments, tribal nations, citizens' groups, and private industry. Absent their involvement and support, it will be difficult to take the steps needed to create an effective and credible stewardship program.

EPA, too, has some responsibility for addressing these issues, as Superfund is the primary statute driving cleanup activities at DOE and many other contaminated sites. The increasingly frequent use of institutional controls as an integral component of site remedies demands that the issues of assuring the long-term integrity of institutional controls be addressed.[56] This could be accomplished by amending the Superfund law, or by revising the major Superfund regulation, the National Contingency Plan.[57]

In moving forward to create a stewardship program, one of the major issues is the scope of such a program. There are a number of types of sites that could be included under the umbrella of a single stewardship program, although different types of sites might warrant slightly different stewardship approaches. These include

• Contaminated DOE sites;

---

[56]See recommendations of Hersh and others, *Linking Land Use.*

[57] *National Oil and Hazardous Substances Pollution Contingency Plan,* 40 CFR § 300.

- Federal facilities on the Superfund National Priorities List (NPL);
- Privately owned Superfund sites;
- Hazardous waste management facilities regulated under Subtitle C of the Resource Conservation and Recovery Act (RCRA);[58] and
- Closed commercial nuclear power facilities.

Arguably some kind of post-closure care, or long-term stewardship, will be required at all of these types of sites. Many federal and private sites addressed under the Superfund and RCRA programs will require stewardship, whether this takes the form of inspecting a site to ensure that institutional controls are being complied with, or monitoring the site to make sure that the integrity of the remedy is maintained.

The question of the appropriate scope of a stewardship program is important, as the decision of whether to create a stewardship program for DOE sites (only), or for some broader set of contaminated sites, affects what federal agency or department should take the lead in moving a program forward, what kind of authorizing legislation makes sense, what organization should be charged with external oversight, and the level of support in Congress for creating such a program, among other issues.[59] Because all of the above types of sites, except commercial nuclear facilities, are already subject to Superfund and RCRA, we confine our recommendations to those sites already regulated under these laws and do not address stewardship at commercial nuclear facilities in this report.[60]

The issue of what agency should conduct independent oversight of stewardship is an important one. "External regulation" of DOE, long a hot topic in the nuclear safety arena, is critical to a credible stewardship program. Even with the strides made over the past few years, DOE still does not enjoy a high degree of public trust and confidence. This means that either EPA, the Nuclear Regulatory Commission, or some other organization will need to be given new oversight responsibilities for DOE sites. No matter which organization is given this charge, some major internal changes will be needed if it is to successfully (and credibly)

---

[58] *Resource Conservation and Recovery Act of 1976,* Public Law 580, 94th Cong., 2nd Sess. (October 21, 1976), § 3001–3023.

[59] For a summary of RFF's 1998 workshop on long-term stewardship, which addressed many of these issues, see RFF's web page at *http://www.rff.org/conf_workshops/files/stewardship98.htm.*

[60] This is not to say that such a program is not needed, but simply that we do not address this issue here.

perform this new function. A wide-ranging debate, such as has occurred regarding external regulation of DOE nuclear safety, needs to take place.

Recommendation

There are a number of steps that should be taken by DOE, EPA, and Congress to help move this important issue forward. We have grouped these steps into two categories: (1) steps that can be taken now to create a stewardship program, and (2) research and analysis that should be conducted to help support the development of a stewardship program. Needless to say, there should be early, meaningful, and continual involvement of all stakeholders in the development of a stewardship program.

Creating a Stewardship Program

A number of steps can and should be taken to create a stewardship program. We list here four major recommendations in their desired order, beginning with legislative changes that are needed and ending with administrative actions. If Congress fails to address stewardship legislatively, EPA and DOE should move forward to address this issue administratively. These recommendations are not mutually exclusive, although it is probably not necessary to implement all of them at the same time. For example, if Congress were to enact stewardship legislation, then it would not be necessary to bring together a group of stakeholders to discuss what kind of legislation would be necessary. Such a group might still be needed, however, to help develop effective stewardship policies.

1. Congress should enact stand-alone stewardship legislation or amend Superfund and RCRA to include specific language requiring the creation of a stewardship program for all contaminated sites requiring post-closure care that are regulated under the nation's environmental laws. These amendments should delineate the specific functions that comprise stewardship, assign these responsibilities to specific agencies or departments (including states), and create formal mechanisms for external accountability. The law should also require federal and state environmental agencies and the regulated community to estimate the full costs of post-closure care, and to specify what organization is responsible for conducting post-closure (or stewardship) activities, and for paying for these activities. In addition, Congress should require the federal government to

develop a national registry of all federal and private sites addressed under Superfund and RCRA that require the use of institutional controls.

2. EPA should amend the National Contingency Plan to clearly define post-closure responsibilities at Superfund sites on the part of federal, state, and local governments, and regulated entities—both public and private. The revised NCP should clarify requirements for the use and enforcement of institutional controls at Superfund sites, including federal facilities. EPA should promulgate similar regulations for sites regulated under Subtitle C of RCRA.

3. The President's Council of Environmental Quality, jointly with EPA, should convene an interagency task force that also includes independent experts and representatives of major stakeholders to develop a government-wide policy on long-term stewardship at both federal and private sites regulated under Superfund and RCRA. This task force should be charged with making specific recommendations regarding the structure of a stewardship program for federal and private sites, outline the need for a legislative stewardship mandate, and address the appropriate roles and responsibilities of different levels of government for post-closure care at contaminated sites.

4. The secretary of DOE should create a high-level task force that includes independent experts from the full panoply of stakeholders—federal, state, and local governments; tribal nations; environmental groups; private industry; and academia—to develop a stewardship mission for DOE, and make specific recommendations for integrating the costs and challenges of long-term stewardship into the major DOE internal decision-making and budgeting processes, including site-specific plans and records of decision. In the interim, DOE should create an office of long-term stewardship to develop and coordinate stewardship policies, conduct research and other activities, and craft its own internal long-term stewardship policy. This office should report directly to the secretary, at least in the first few years, as it seeks to coordinate policies and programs across all DOE offices.

Stewardship Research and Analysis

Whether or not a legislative mandate for stewardship is enacted now or years from now, better information on the scope, nature, and cost of a stewardship program is needed.

1. EPA should commission two studies regarding stewardship at both federal and private sites: (a) an examination of the role of states and local governments as stewardship implementors and/or overseers, and (b) an evaluation of a range of institutional alternatives for assuring long-term compliance with institutional controls.

2. EPA should assess the need for long-term stewardship at sites addressed under Superfund and Subtitle C of RCRA, and estimate the full cost of stewardship to both the public and private sectors at these sites.

3. DOE should (a) estimate the cost, time frames, and types of activities that will be needed for long-term stewardship at the sites in the weapons complex; and (b) commission an independent report regarding the appropriate role of tribal nations and local governments in long-term stewardship at DOE sites.

4. The Congressional Budget Office should conduct a study of alternative funding schemes, for both federal and private sites, for paying for long-term stewardship. This study should examine the viability of a federal trust fund, state trusts, and private trusts, as well as other possible funding mechanisms.

5. The Congressional Research Service or National Academy of Sciences should conduct a study of how other countries are meeting their stewardship responsibilities for post-closure care at contaminated sites to help inform the development of a U.S. program.

Appendices

Appendix A. Federal Environmental Programs with Stewardship Elements

PROGRAM	Land Disposal of Hazardous Waste	Sites with Hazardous and/or Radioactive Contamination	Comm Low-1 Radio Waste I
PROGRAM ATTRIBUTE			
Relevant Statute(s)	RCRA	CERCLA, RCRA	NW AE
Major Regulation(s)	40 CFR 264	40 CFR 300, 40 CFR 264	10 CF
Promulgator of Regulation(s)	EPA, states	EPA, states	NR agreeme
Type of Waste/Contamination	Hazardous waste	Radioactive and hazardous contamination	Low- radio wa
Period of Hazard Duration (Years)	Perpetuity, in some cases	Perpetuity, in some cases	In ge less than
Regulated Unit	Hazardous waste disposal facilities	Site as defined by EPA	Low-lev disposal
Duration Design Requirements for Regulated Unit	No time period specified	No time period specified	500-yea of pro
Regulatory Protection Standards (mrem Per Year or Life-Time Risk)[a]	Variable	Life-time risk of $10^{-4} - 10^{-6}$	25 mrem maxi permissible

Note: [a]mrem = millirem (one-thousandth of a rem). A rem (Roentgen Equivalent Man) is a measurement of the effect of radiation on human tissue, taking into account both the amount of energy absorbed by the tissue and the relative biological damage caused by the type of radiation absorbed.

...missioned ...mercial ...uclear ...cilities	Uranium Mill Tailings Sites	Geologic Repository for Spent Nuclear Fuel & High-Level Waste	Geologic Repository for Transuranic Wastes
...AEA	UMTRCA	NWPA	WIPP Authorization and Land Withdrawal Act
...CFR 20	40 CFR 192, 10 CFR 40	10 CFR 60, 40 CFR 197 (to be issued)	40 CFR 191, 40 CFR 194
...NRC, ...nent states	EPA, NRC	NRC, EPA	EPA
...ioactive ...mination	Uranium mill tailings	High-level waste, spent nuclear fuel	Transuranic waste
...general, ...1,000 years	Radium: 16,000 years	Focus is on first 10,000 years	Plutonium: 240,000 years
...erly used ...d buildings	Uranium mill tailings disposal facilities, contaminated soil and groundwater	High-level waste geologic repository	Transuranic waste disposal facility and surrounding areas
...00 years	200 – 1,000 years	Not yet promulgated	10,000 years
...em per year ...ximum ...ible exposure	Life-time risk of $10^{-3} – 10^{-4}$	Not yet promulgated	15 mrem per year maximum permissible exposure

Appendix A. Federal Environmental Programs with Stewardship Elements–continued

PROGRAM STEWARDSHIP ELEMENT	Land Disposal of Hazardous Waste	Sites with Hazardous and/or Radioactive Contamination	Comme Low-L Radioa Waste Di
Institution(s) Charged with Post-Closure Program Regulation	EPA, states	EPA, state, & local governments	NRC agreemen
Institution(s) Charged with Post-Closure Program Implementation	Facility operator	EPA, state, and/or responsible parties	Federa state ag
Regulatory Time Frame (Years) for Post-Closure Care	30 years, but administrator has discretion	Perpetual; five-year review requirements	100 years n after clc
Post-Closure Monitoring and Maintenance	Yes	Yes, if released for restricted use	Yes
Institutional and Other Controls (as needed)	Yes	Yes	Yes
Information Management	Yes	Yes	Yes
Specific Post-Closure Funding	Yes, funding provided by permittee	Perpetual liability of responsible parties or Superfund	Yes funding p by lice

Note: bPrimary source of funds is a fee on utilities, which is passed on to their customers.

missioned mercial clear ilities	Uranium Mill Tailings Sites	Geologic Repository for Spent Nuclear Fuel & High-Level Waste	Geologic Repository for Transuranic Wastes
RC, ent states	NRC	NRC	EPA
third party, ment entity	DOE	DOE	DOE
,000 years, ne cases, jective but quirement	Indefinitely	To be determined	100 years minimum after closure
eleased for icted use	Yes, for Title II sites	Yes	Yes
Yes	Yes	Yes	Yes
atus report quired	Yes	Yes	Yes
Yes, provided by if needed	Yes, financial surety requirements apply to new licensees	Yes, Nuclear Waste Disposal Fund[b]	No

Appendix B. DOE Sites with Current EM Activity: FY 1998 EM Appropriations

State	Site	$ (thousands)
Major Nuclear Weapons Production Sites (5)		
South Carolina	Savannah River Site	1,158,744
Washington	Hanford Site	1,067,740
Colorado	Rocky Flats Environmental Technology Site	632,100
Idaho	Idaho National Engineering and Environmental Laboratory	440,910
Tennessee	Oak Ridge Reservation (all sites)	387,330
Other Nuclear Weapons Production Sites (9)		
Ohio	Fernald Environmental Management Project	258,700
Ohio	Miamisburg Environmental Management Project	86,622
Missouri	Weldon Spring Site	65,800
Ohio	Portsmouth Gaseous Diffusion Plant	45,502
Kentucky	Paducah Gaseous Diffusion Plant	43,054
Texas	Pantex Plant	24,541
Colorado	Grand Junction Projects Office Site	16,412
Ohio	Ashtabula Environmental Management Project	14,710
Missouri	Kansas City Plant	4,522
Active DOE Research Labs (14)		
New Mexico	Los Alamos National Laboratory	128,957
California	Lawrence Livermore National Laboratory Main Site	54,543
California	Lawrence Livermore National Laboratory Site 300	a
New Mexico	Sandia National Laboratories - New Mexico	45,190
California	Sandia National Laboratories - California	a
New York	Brookhaven National Laboratory	24,900
Illinois	Argonne National Laboratory - East	16,319
California	Lawrence Berkeley National Laboratory	11,177
Idaho	Argonne National Laboratory - West	3,600
New Jersey	Princeton Plasma Physics Laboratory	3,389
California	Stanford Linear Accelerator Center	995
New Mexico	Lovelace Respiratory Research Institute	743
Iowa	Ames Laboratory	363
New York	Separations Process Research Unit	0[b]
Research Labs with No Future DOE Missions (5)		
California	Energy Technology Engineering Center	17,426
Ohio	Columbus Environmental Management Project - King Avenue	12,494
Ohio	Columbus Environmental Management Project - West Jefferson	a
California	Laboratory for Energy-Related Health Research	5,156
Puerto Rico	Center for Energy and Environmental Research	c

State	Site	$ (thousands)
Testing Sites (10)		
Nevada	Nevada Test Site	60,126
Nevada	Central Nevada Test Site	3,006
Mississippi	Salmon Site	2,270
New Mexico	Gnome-Coach	971
Colorado	Rulison	929
Alaska	Amchitka Island	911
Nevada	Shoal Site	723
New Mexico	Gasbuggy	397
Colorado	Rio Blanco	262
Nevada	Tonopah Test Range Area	c
Uranium Mill Processing Sites (5)		
Utah	Monticello Remedial Action Project	23,616
North Dakota	Belfield	d
North Dakota	Bowman	d
Colorado	Maybell	d
Colorado	Naturita	d
Other EM Sites (5)		
New Mexico	Waste Isolation Pilot Plant	194,866
New York	West Valley Demonstration Project	114,256
California	General Atomics Site	4,100
California	General Electric Vallecitos Nuclear Center	106
Kentucky	Maxey Flats Disposal Site	e

Source: Office of Environmental Management, "Environmental Management Program Budget Totals (FY 1997 - FY 1999)" in *The FY 1999 Environmental Management Budget Request: Budget Request Presentation* (Washington, D.C.: U.S. Department of Energy, January 31, 1998).

Notes: Dollar amounts represent site-specific project and privatization funds only. Science & Technology and Program Direction funds excluded.

a. The Columbus EM Project, Lawrence Livermore National Laboratory, and Sandia National Laboratories each have two listings in *Accelerating Cleanup*, but are consolidated in DOE budget documents.

b. EM funding of decontamination activities at the Separations Process Research Unit (part of the Knowles Atomic Power Laboratory) are not expected to begin until FY 2000.

c. Projects at the Tonopah Test Range Area and the Center for Energy and Environmental Research are funded by the Nevada and Oak Ridge Operations Offices, respectively.

d. Individual site breakdowns of UMTRA funding are not available.

e. The Albuquerque Operations office funds approximately 40% of the Superfund cleanup costs at the state-owned Maxey Flats low-level waste disposal facility.

Appendix C. DOE Activities at EM Sites by Program Office: FY 1998

SITE	Environmental Management
Amchitka Island	●
Ames Laboratory	●
Argonne National Laboratory - East	●
Argonne National Laboratory - West	●
Ashtabula Environmental Management Project	●
Belfield	●
Bowman	●
Brookhaven National Laboratory	●
Center for Energy and Environmental Research	●
Central Nevada Test Site	●
Columbus Environmental Management Project - King Avenue	●
Columbus Environmental Management Project - West Jefferson	●
Energy Technology Engineering Center	●
Fernald Environmental Management Project	●
Gasbuggy	●
General Atomics Site	●
General Electric Vallecitos Nuclear Center	●
Gnome-Coach	●
Grand Junction Projects Office Site	●
Hanford Site	●
Idaho National Engineering and Environmental Laboratory	●
Kansas City Plant	●
Laboratory for Energy-Related Health Research	●
Lawrence Berkeley National Laboratory	●
Lawrence Livermore National Laboratory Main Site	●
Lawrence Livermore National Laboratory Site 300	●
Los Alamos National Laboratory	●
Lovelace Respiratory Research Institute	●

...se ...ms	Energy Research	Nuclear Energy	Other
	●		●
	●	●	●
		●	
	●	●	●
	●		
	●		
	●	●	●
	●	●	●
			●
	●		●
	●	●	●
	●	●	●
	●		

Appendix C. DOE Activities at EM Sites by Program Office: FY 1998–continued

SITE	Environment Management
Maxey Flats Disposal Site	●
Maybell	●
Miamisburg Environmental Management Project	●
Monticello Remedial Action Project	●
Naturita	●
Nevada Test Site	●
Oak Ridge Reservation (all sites)	●
Paducah Gaseous Diffusion Plant	●
Pantex Plant	●
Portsmouth Gaseous Diffusion Plant	●
Princeton Plasma Physics Laboratory	●
Rio Blanco	●
Rocky Flats Environmental Technology Site	●
Rulison	●
Salmon Site	●
Sandia National Laboratories - California	●
Sandia National Laboratories - New Mexico	●
Savannah River Site	●
Separations Process Research Unit	
Shoal Site	●
Stanford Linear Accelerator Center	●
Tonopah Test Range Area	●
Waste Isolation Pilot Plant	●
Weldon Spring Site	●
West Valley Demonstration Project	●

Source: U.S. Department of Energy, "FY 1999 Laboratory/Facility Table," in *Department of Energy FY 1999 Congressional Budget Request* (Washington, D.C., February 1998).

Notes: "●" indicates that funds were appropriated to the program office for activities at the site.

se ns	Energy Research	Nuclear Energy	Other
		•	
			•
	•	•	•
		•	
			•
		•	
	•		•
			•
	•	•	•
	•		•
	•	•	•
		•	
	•		

Program offices represented in "other" include: Energy Efficiency and Renewable Energy (EE); Nonproliferation and National Security (NN); Fossil Energy (FE); and Fissile Materials Disposition (MD).

Appendix D. Site Missions at Major EM Sites

DOE OFFICE	Office of Defense Programs	Office o Energy Res
DOE SITE		
Oak Ridge Reservation	Weapons component fabrication ● HEU strategic reserve storage	Energy rese laboratorie Particle acceler Fusion rese
Savannah River Site	Tritium recycling ● Possible tritium production	No know activity
Idaho National Engineering and Environmental Lab	Various nonnuclear national security activities	Energy rese laboratorie Fusion rese
Rocky Flats Environmental Technology Site	No known activity	No know activity
Hanford	Possible tritium production at Fast Flux Test Facility	Energy rese labs

Sources: Office of Resource Management and Services, *Field Facility Fact Book: Field Management* (U.S. Department of Energy, November 1996); U.S. Department of Energy, *Storage and Disposition of Weapons Usable Fissile Materials: Final Programmatic Environmental Impact Statement* (Washington, D.C., December 1996); U.S. Department of Energy, *Final Programmatic Environmental Impact Statement for Stockpile Stewardship and Management* (Washington, D.C., September 1996); U.S. Department of Energy, *Charting the Course: The Future Use Report,* DOE/EM-0283 (Washington, D.C., April

Nuclear Science, hnology	Office of Fissile Materials Disposition	Other Activities	Ultimate Site Disposition (Based on Future Use Report, BEMR and DOE Web Sites) [a]
eactor: high e reactor ● e nuclear ɔr the NRC	Surplus HEU[b] blending and disposition	Genome research ● Materials research ● Life Sciences and Environmental Technologies Lab ● Center for Manufacturing Technology	Mixed use, but to remain an important national laboratory ● Continued defense programs ● Some areas non-DOE industrial and commercial use
nown ivity	Surplus HEU[b] blending and disposition ● Possible plutonium disposition activities	Savannah River Technology Center ● Savannah River Ecology Lab	Mixed use with continued defense program activity
ɪ reactor: test reactor ting facilities	Possible plutonium disposition activities	National Environmental Engineering and Technology Center	Mixed use with continued operations of INEEL as a national lab
nown ivity	No known activity	No known activity	Site closure—no DOE missions after cleanup is complete ● Non-DOE research and industrial activity
nal Nuclear gram (to aid t Union n Europe)	Possible plutonium disposition activities	Environmental Molecular Sciences Laboratory	Mixed use, including both DOE and non-DOE research and industrial activity

1996); Office of Environmental Management, *The 1996 Baseline Environmental Management Report*, DOE/EM-0290 (Washington, D.C.: U.S. Department of Energy, June 1996); various DOE Budget Documents and World Wide Web sites.

Notes: [a]BEMR = Baseline Environmental Management Report.
[b]HEU = highly enriched uranium.

Resources for the Future is an independent nonprofit organization engaged in research and public education with issues concerning natural resources and the environment. Established in 1952, RFF provides knowledge that will help people to make better decisions about the conservation and use of such resources and the preservation of environmental quality.

RFF has pioneered the extension and sharpening of methods of economic analysis to meet the special needs of the fields of natural resources and the environment. Its scholars analyze issues involving forests, water, energy, minerals, transportation, sustainable development, and air pollution. They also examine, from the perspectives of economics and other disciplines, such topics as government regulation, risk, ecosystems and biodiversity, climate, Superfund, technology, and outer space.

Through the work of its scholars, RFF provides independent analysis to decisionmakers and the public. It publishes the findings of their research as books and in other formats, and communicates their work through conferences, seminars, workshops, and briefings. In serving as a source of new ideas and as an honest broker on matters of policy and governance, RFF is committed to elevating the public debate about natural resources and the environment.

Milton Keynes UK
Ingram Content Group UK Ltd.
UKHW031134141024
449569UK00006B/206